Collaborating for Digital Transformation

Collaborating for Digital Transformation

How Internal and External Collaboration Can Contribute to Innovate Public Service Delivery

Edited by

Koen Verhoest

Professor of Public Governance and Administration, Department of Political Science and GOVTRUST Centre of Excellence, University of Antwerp, Belgium

Gerhard Hammerschmid

Professor of Public and Financial Management, Hertie School, Germany

Lise H. Rykkja

Professor of Political Science/Public Administration, Department of Government, University of Bergen, Norway

Erik Hans Klijn

Professor of Public Administration, Department of Public Administration, Erasmus University Rotterdam, the Netherlands

EE Edward Elgar PUBLISHING

Cheltenham, UK • Northampton, MA, USA

Published by
Edward Elgar Publishing Limited
The Lypiatts
15 Lansdown Road
Cheltenham
Glos GL50 2JA
UK

Edward Elgar Publishing, Inc.
William Pratt House
9 Dewey Court
Northampton
Massachusetts 01060
USA

A catalogue record for this book
is available from the British Library

Library of Congress Control Number: 2023949799

This book is available electronically in the Elgaronline
Political Science and Public Policy subject collection
http://dx.doi.org/10.4337/9781803923895

MIX
Paper from
responsible sources
FSC® C013056

ISBN 978 1 80392 388 8 (cased)
ISBN 978 1 80392 389 5 (eBook)

Printed and bound in Great Britain by
TJ Books Limited, Padstow, Cornwall

Contents

Figures

Tables

Contributors

Jessica Breaugh is Lecturer of Corporate Social Responsibility and Business Ethics at ESCP, and former Post Doctoral scholar at the Centre for Digital Governance at the Hertie School in Berlin, Germany. Her research primarily focuses on the study of motivation and behaviour of public servants (such as collaborative work), and the meaning of work in the 21st century. Given her teaching focus, she is increasingly interested in the topics of sustainability and the overlapping spheres of public and private responsibilities. Jessica has published several papers in leading public management journals.

Lena Brogaard is Associate Professor in Public Administration at Roskilde University, Denmark. Her research focuses on competitive and collaborative approaches to involving businesses in the production and delivery of public services. Lena has published in high-ranking international journals on innovation and other outcomes from cross-sector partnerships and public procurement.

Chesney Callens is a Postdoctoral Research Fellow within the research group of Politics & Public Governance (Department of Political Sciences) at the University of Antwerp, Belgium. He holds a master's degree in Public Administration and Management from the University of Ghent (Belgium). His research interests are collaborative and network governance, innovation management, collaborative innovation and structural reforms.

Jaime García-Rayado is Assistant Professor of Accounting and Finance at the University of Zaragoza, Spain. His research focuses on public sector innovation, e-government and co-production of public services. Jaime has published in international scientific journals, such as *Government Information Quarterly* and *Profesional de la información.*

Gerhard Hammerschmid is Professor of Public and Financial Management at the Hertie School in Berlin, Germany. He is Founding Director of the School's Centre for Digital Governance and his research focuses on public management reform, performance management and comparative public administration, with a particular emphasis on the dynamics and impact of digitalization in a government context. Gerhard contributed to various EU-funded research projects – most recently the EU Horizon H2020 TROPICO project where he led Work Package 6 – and his research has been published in international journals, such as *Public Administration Review*, *Government Information Quarterly*, *Governance* and *Public Administration.*

Erik Hans Klijn is Professor at the Department of Public Administration and Sociology at the Erasmus School of Social and Behavioural Studies, Erasmus University Rotterdam, the Netherlands. His research is about network governance, network management, but also about the influence of media on complex decision-making and public branding.

Tom Langbroek is a PhD candidate at the Politics and Public Governance research group at the University of Antwerp, Belgium. He is interested in issues regarding inter-organizational collaboration to enhance government functioning in innovative ways. Alongside his academic activities, he is a senior policy officer for 'COVID-19 Information and Coordination' for the Dutch Ministry of Health, Welfare and Sports.

Veiko Lember is a Senior Research Fellow in Public Management and Policy at Tallinn University of Technology and Visiting Professor at KU Leuven, Belgium. His main research interests are in public administration, with a particular emphasis on the intersection of technology, innovation and the state. His most recent book with Tiina Randma-Liiv is *Engaging Citizens in Policy Making: e-Participation Practices in Europe* (2022, Edward Elgar).

Miriam Lips is Professor of Digital Government at Victoria University of Wellington, New Zealand. Her research primarily focuses on digital government and democracy. Miriam has published extensively in this field, with her latest book entitled *Digital Government: Managing Public Sector Reform in the Digital Era* (2020, Routledge).

Steven Nõmmik is an Early Stage Researcher at the Ragnar Nurkse Department of Innovation and Governance at Tallinn University of Technology in Tallinn (TalTech), Estonia. His research interests are connected to public management, coordination, innovation studies with a focus on public sector collaboration impact on implementation of digital innovation.

Vicente Pina (PhD) is Professor at the University of Zaragoza, Spain. He was a visiting researcher at the University of Strathclyde, Glasgow (UK). He has published in some of the best public sector management ISI journals – *Public Administration*, *Governance*, *Government Information Quarterly*, *International Public Management Journal*, *Transportation Research Part A*. Vicente has been a partner in TROPICO and ROBUST EU projects. He has also carried out several research contracts with the Spanish Internal Audit Office and various Regional Audit Offices. He has won several international and national research awards from *International Public Management Journal* and the US Academy of Accounting and Financial Studies.

Maike Rackwitz holds a PhD in Governance from the Hertie School in Berlin, Germany, and currently works as a public sector consultant. Her main research interests lie in digital and urban governance, with a particular focus on public sector reform and decision-making in national and local governments. Maike has published on topics related to collaboration, digitalization and corporatization.

Lise H. Rykkja is Professor of Political Science/Public Administration and Head of Department at the Department of Government, University of Bergen, Norway. Her research concentrates on the organization and development of public administration and democratic institutions, administrative policies and reforms with a particular focus on problems in the public sector, including crisis management and digitalization. Lise coordinated the EU Horizon 2020 project Transforming into Open, Innovative and Collaborative Governments (TROPICO).

Dries van Doninck is a doctoral student at the Department of Management, Faculty of Business and Economics and is affiliated to the research group Politics & Public Governance, both at the University of Antwerp, Belgium. His current research focuses on stress in the public sector and the effects of leadership, innovation, organizational change and behaviour on stress and vice versa.

Koen Verhoest is Professor at the University of Antwerp, Belgium, and is coordinating the Politics & Public Governance research group as well as the interdisciplinary Centre of Excellence GOVTRUST. Koen studies autonomy, control, coordination and collaboration in interorganizational relations related to public and regulatory tasks, and their relation with, among others, innovation, reputation and trust. He has published extensively on these issues in high-ranking journals, and in several books published by Edward Elgar (e.g., *Trust in Regulatory Regimes*, 2017) and Palgrave Macmillan.

Abbreviations

Abbreviation	Definition
AI	Artificial Intelligence
BAU	Business as Usual
ESOC	European Space Operations Command
ETCB	Estonian Tax and Customs Board
EU	European Union
fsQCA	Fuzzy-Set Qualitative Comparative Analysis
GaaP	Government as a Platform
GDS	Government Digital Services
ICT	Information and Communication Technology
IMEC	Interuniversity Microelectronics Center
IoT	Internet-of-Things
NGO	Non-Governmental Organisation
NPG	New Public Governance
OAA	Online Access Act
OECD	Organisation for Economic Co-operation and Development
PoC	Proof of Concept
PPI	Public-Private Innovation Partnership
PPP	Public-Private Partnership
QCA	Qualitative Comparative Analysis|
SDG	Single Digital Gateway

TROPICO	Transforming into open, innovative, and collaborative government
TU Darmstadt	Technical University Darmstadt

PART I

Introduction

1. Introduction: collaborating for digital transformation in the public sector

Lise H. Rykkja, Gerhard Hammerschmid, Erik Hans Klijn and Koen Verhoest

INTRODUCTION

Governments around the globe, international organizations, as well as the European Union, recognize the need for digital, open, and collaborative governments. Digital transformation promises a pathway forward for governments and organizations that aim to improve their policy design, services, and efficiency. At the same time, intergovernmental and public-private collaboration is actively, widely, and increasingly used to achieve such transformation. This book investigates both, specifically the interlinkages between digital transformation and collaboration, and how collaboration works – in practice – to support and/or promote the digital transformation our society and governments are going through.

Collaboration within governments and between public and private actors has always played a crucial role in policy development and implementation. Both themes have been covered a lot in research over the last two decades. For collaboration within governments, one can think of the classic work of Agranoff (1986) or Hanf and Scharpf (1978) on which a wide literature has been built. If it comes to collaboration between public and private organizations there is a wide tradition of collaborative and network governance that has dealt with this over the past two decades (see, e.g., Kickert et al., 1997; Huxham and Vangen, 2005; Ansell and Gash, 2008; Emerson et al., 2012). However, combining these traditions with the literature and practice of digital transformation is understudied.

Our book aims to contribute more specifically to understanding how governments move towards digital transformation and interact with others in this collaborative endeavour. It theoretically develops and empirically explores the interlinked aspects of digital transformation, intergovernmental collaboration, collaborative governance, and public sector innovation through dedicated and diverse chapters. It does so by connecting the long-standing literature on collaboration, and collaborative and network governance with literature about digitalization and digital transformation and applying this to a large number of cases throughout European Union (EU) countries.

It thus provides a strategic perspective to the governance of digital transformation. This book explicitly explores the relevance of the digital transformation of government and the interplay between digital transformation and governance. It provides a brief introduction to the topic of intergovernmental and public-private collaboration to achieve digital transformation, addresses key challenges of collaboration, and more specifically looks at how institutional design and leadership can help overcome such challenges and foster collaborative dynamics to achieve digital transformation. The general aim is to provide significant insights to both academics and practitioners through a large variety of research methodologies and a comparative analysis of a diverse set of empirical cases across Europe.

The book is a result of intense collaboration and extensive research done in and related to an EU Horizon 2020 research project funded from 2017 to 2021: 'TROPICO: Transforming into open, innovative, and collaborative government'.[1] TROPICO studied *internal collaboration* within the public sector looking at whether and under which conditions collaborative public management delivers innovative public services and advances digital government across levels and between organizations within government. The project furthermore devoted research to practices of *external collaboration*, studying different kinds of partnerships between government, private stakeholders, and users in the creation and delivery of innovative and digital public services.

Most of the chapters in this book draw directly from research within the project. An additional chapter from Miriam Lips puts this research into a broader context of international research findings on digital transformation in government.

DIGITAL TRANSFORMATION AND COLLABORATION

The concept of digital transformation has injected new vigour into both policy-making and conceptual and empirical research across the world. Governments see digital transformation as an opportunity to improve service delivery and policies, change or streamline public sector operations, processes, and design, and collaborate in new ways, creating more transparency, interoperability, or citizen satisfaction (Curtis, 2019; Mergel, Edelmann, and Haug, 2019). The European Commission recognizes the urgency and relevance of the 'Digital Decade',[2] as the COVID-19 pandemic has made the need for digitization in our governments and societies abundantly clear. Thanks to digital technologies, societies were able to mitigate many of the detrimental effects on the economy and society of the COVID-19 crisis. This would not have been possible only a few decades ago. The European Commission (2021) prioritizes a strategic orientation towards digital transformation and has recently formulated a Digital Compass, which connects four cardinal points of digitalization: (1) A digitally skilled population and highly skilled digital professionals; (2) Secure and performant sustainable digital infrastructures; (3) Digital transformation of businesses; and (4) Digitalization of public services. For this book, particularly the fourth cardinal point is of interest, as it pertains to how governments in Europe can digitally transform their services.

There is a myriad of approaches to and definitions of digital transformation (Vial, 2019). Diverse initiatives to promote the digital transformation of public services have been launched widely in many countries and are used to initiate various strategizing and organizing activities. The European Commission, for example, has adopted a strategy aimed at 'A Europe fit for the digital age', aiming to empower people and businesses to make use of new technology.[3] However, many authors point to the fact that a universal and comprehensive understanding of the concept is lacking (Gong and Ribiere, 2021). Mergel et al. (2019) focus on the transformational part of the concept, highlighting the evolvement and outcomes of digital transformation and seeing it as a holistic effort to revise core processes and services of government, evolving along a continuum of transition from analogue to digital in the process of reviewing and revising current policies, processes, and services. Gong and Ribiere (2021, p. 12) seem to capture some of the essence in their unified definition of digital transformation as: 'A fundamental change process, enabled by the innovative use of digital technologies accompanied by the strategic leverage of key resources and capabilities, aiming to radically improve an entity* and redefine its value proposition for its stakeholders.' (*An entity could be: an organization, a business network, an industry, or society.) Similarly, Hinings, Gegenhuber, and Suddaby (2018, p. 52) from an institutional organization theory perspective understand digital transformation as 'the combined effects of several digital innovations bringing about novel actors (and actor constellations), structures, practices, values, and beliefs that change, threaten, replace or complement existing rules of the game within organizations and fields'.

Related to the many definitions, digital transformation is achieved in many ways. A pure technological approach may provide some of the necessary tools and building blocks, but digital transformation is also about creating an environment for change to happen. Here, also managerial, organizational, and processual factors are crucial. Thus, digital transformation requires adequate governance structures, management, leadership, information sharing, cross-organizational collaboration, specific skills, expertise, and mindsets.

In the literature, the digital transformation of services goes beyond the simple development and implementation of information and communication technology (ICT)-enabled services. On a macro level, digital transformation hence relates to the profound changes societies, industries, and governments undergo because of the introduction of digital technologies (Vial, 2019). On an organizational level, digital transformation corresponds to the organizational strategy or trajectory towards change and transformation by using digital technologies (Singh and Hess, 2017). Digital transformation at the macro level can act as a contextual feature, which necessitates the introduction of digital transformation strategies at an organizational level but also implies the ability to innovate and implement new digital solutions.

In the government's pursuit of digital transformation, its ability to achieve ICT-enabled services in collaboration with internal and external stakeholders, such as government agencies, market actors, third-sector organizations, civil society, citizens, and service users, is crucial. Both internal and external collaboration networks enable the connection between stakeholders' capabilities and help to align the problem-solving efforts of stakeholders, which stimulates technological innovation and increases the impact of digital transformation (Camarinha-Matos et al., 2019). This is also strongly linked to findings from research on collaborative public sector innovation (Torfing, 2019) which emphasizes both the (resource) dependency of actors and their availability of different information (and perspectives) as core elements for innovation.

Ansell and Gash (2008, p. 543) argue that collaboration has emerged as a new form of governance to replace adversarial and managerial modes of policymaking and implementation. This governance form typically brings public and private stakeholders together with public agencies, resulting in more consensus-oriented decision-making. The ensuing research has found that there exist a large variety of approaches to collaboration and that the term is used to describe a number of different practices, relating to participation, agency, inclusiveness, and scope – and based on different normative assumptions (Batory and Svensson, 2018). As society and governments face more complex cross-cutting policy challenges, calls for better coordination of government policies and more transboundary collaboration also within government, between multiple levels of government, and in public administration, in general, have become stronger (Lægreid et al., 2014; Molenveld et al., 2020; Molenveld, Verhoest, and Wynen, 2021; Lægreid and Rykkja, 2022). Collaborative frameworks, cutting across organizational boundaries, are frequently seen as challenging the traditional hierarchical 'silo' structures and closed top-down processes that characterize public organizations (Scott, 2020), resulting in more innovation (Bommert, 2010; Callens et al., 2022). A move away from New Public Management (NPM)-driven disaggregation towards reintegration and needs-based holism which imply whole-of-government solutions and collaboration is regarded as a key characteristic of a 'digital-era governance' (Dunleavy et al., 2006).

So, the literature suggests that multi-actor collaboration can spur public innovation and hence move digital transformation forward. Collaboration is essential because it brings actors with different world views, experiences, and innovation assets (like information and resources) together. This can potentially create productive destructions of existing mindsets and perceptions and trigger the formulation of new creative ideas, pool the resources and capacities needed to transform these ideas into innovative products, organizational designs, and procedures, and promote innovation diffusion to relevant audiences (Agger and Sørensen, 2018, p. 54). However, collaboration within and between governments, and between governments and external stakeholders (e.g., firms, non-profits, citizens) is far from obvious, as new challenges may arise in the form of network complexities, power imbalances, and increased risks (Huxham and Vangen, 2005; Klijn and Koppenjan, 2016; Torfing, 2019). Collaboration within digitalization projects is even more intricate because of the variety of stakeholders involved, as well as the content, resources, and (technical) knowledge required to be successful (Picazo-Vela et al., 2018; Callens, 2022; Dockx et al., 2022; Breaugh, Rackwitz, and Hammerschmid, 2023).

Despite the growth and interest in collaborative innovation within the realm of digitalization, there are still many unanswered questions regarding how public organizations should govern and organize collaborative processes within digital projects (Lopes and Farias, 2020). To achieve digital transformation, collaboration networks need to be properly *designed* (in terms of their organizational arrangements, process agreements, formal regulations, the constellation of engaged stakeholders, etc.), and *led* (in terms of the type of leadership the coordinators use) (e.g., Torfing, 2019). User involvement and participation are also important topics, as well as assessing the impact of collaboration in terms of achieving expected outcomes related to, for example, more efficiency, better services, or improved policies.

This book addresses many of these aspects and examines intergovernmental collaboration networks as well as collaboration in public-private networks in the context of government digitalization. By connecting insights from digital transformation and collaborative governance, we aim to provide an integrated approach to the digital transformation of public services in a complex and interconnected government and society.

ABOUT THE BOOK

In the following chapters, we explore two major themes of collaborating for digital transformation. First, how do *intergovernmental collaboration* networks stimulate digital transformation in public services? The respective chapters consider the dynamics and challenges of intergovernmental collaboration and study how the institutional design and leadership of intergovernmental networks can promote the development of digital government services and smart cities. Second, the book highlights the relevance of *external collaboration*, looking at how public-private collaboration networks promote the digital transformation of healthcare services. Here, the authors consider how various features of public-private healthcare collaborations – including their structure, management, leadership, and user involvement – stimulate the adoption of innovative eHealth technologies.

The book addresses the two themes through an analysis of empirical data from a total of eight countries within five European administrative traditions: a Nordic tradition (represented by Denmark), a Central and Eastern European tradition (represented by Estonia), a Continental tradition (represented by Germany and the Netherlands), a Napoleonic tradition (represented by Belgium, Italy, and Spain), and the Anglo-Saxon tradition (here, the United Kingdom) (Pollitt and Bouckaert, 2017). The different chapters draw on specific and comparative case studies in these eight countries, including studies of the digital transformation of administrative services of core governments (e.g., online service platforms), smart cities, and healthcare services (e.g., data-sharing platforms, eHealth apps).

Most of the chapters build on extensive and in-depth empirical research. In TROPICO, a total of 29 collaboration projects were studied between 2019 and 2020. Questionnaires and interviews were conducted with 192 practitioners, including project coordinators, senior government officials, executives from private companies, programme managers, representatives of user organizations, and individual users. The TROPICO cases were selected through rigorous case selection criteria. Each case of intergovernmental collaboration was either a key national project used to implement digital platforms in line with the EU Single Digital Gateway regulation or a city with a population of at least 50,000 that was considered a frontrunner in digitalization and was implementing a smart city strategy. Only cases where the main implementing actors could be assigned to the public sector were chosen. Furthermore, only cases that had moved beyond their planning phase were selected. The public-private collaboration networks were selected based on five criteria. First, all cases were established as a formal partnership between public and private actors. Second, each of the cases exhibited some level of user involvement. Third, only cases that implemented or at least thoroughly tested the created services were selected. Fourth, only partnerships that created eHealth services were selected. Fifth, the created eHealth services had to be relatively recent (created within the last five years).

Various methodologies were used to collect and analyse the data in a systematic and comparative form. Data from the intergovernmental collaboration networks was collected and analysed through qualitative case studies and interview coding; 62 semi-structured interviews were carried out for the ten cases in five countries. Data from the public-private collaboration networks was collected and analysed through qualitative case studies, qualitative comparative analysis (QCA), and Q-methodology. A total of 130 interviews, 124 surveys, and 112 Q-sorts were conducted in 19 cases. Research teams from each of the seven European countries were involved in the data collection and analysis of the 29 cases in the book. Besides providing the raw data, the national research teams each also provided detailed case study descriptions of 'their' case, and which also can be found on the TROPICO online case repository.[4]

This systematic and standardized collection of data from many cases contributes to literature and practice, as it transcends the usual single country/single case study/single sector designs, and combines insights from both intergovernmental and public-private collaboration networks on digital transformation. The multi-method approach used in this book, which relies on a variety of data gathering and analysis methodologies, contributes to the empirical rigour and relevance of the results for both scholars and practitioners. The in-depth qualitative analyses also contribute to the formulation of rich insights that should be useful for both project coordinators and policymakers.

Our book aims to be both academic and practical in nature. By combining insights from collaborative governance and collaborative innovation literature with literature on digital transformation, key antecedents of digital transformation in government are suggested and tested. Tangible conditions of enhanced digital transformation related to institutional design, management, and leadership are systematically examined in chapters showcasing findings from in-depth research in two types of collaboration networks: intergovernmental collaborations and public-private collaborations. The integration of various fields of inquiry and the broad scope of types of collaboration networks is intended to contribute to the practical relevance of the book and generates insights that are not restricted to one specific case, service, country, or sector. This rigorous empirical analysis of many cases allows for the formulation of widely applicable lessons for project coordinators who are engaged in similar collaboration networks, and for policymakers that want to exploit the advantages of collaborative governance/innovation and digital transformation.

THE STRUCTURE OF THE BOOK

The book is structured into three parts. The first part (Chapters 2–4) focuses on the conceptual understanding of digital transformation and a theoretical development of the relevance and conditions of intergovernmental and public-private collaboration for digital transformation.

Chapter 2 by Miriam Lips presents an overview of the literature on digital transformation in government and explores the conceptual background of digital transformation. The chapter develops a theoretical perspective for studying digital transformation processes in the public sector, explores empirical research, and provides valuable insights into the conditions that stimulate and influence digital transformation in governments with a special look at digital transformation experiences in New Zealand.

Chapter 3 by Hammerschmid, Breaugh, and Rackwitz presents a conceptual framework for studying intergovernmental collaboration for ICT-enabled services. Drawing on an in-depth literature review of the collaborative and digital governance literature it assesses crucial contextual starting conditions and identifies key collaborative challenges and governance interventions focusing on institutional design and leadership.

In Chapter 4, Callens and Verhoest focus on how collaborative arrangements between public and private actors develop digital service innovations. The chapter presents a theoretical framework that aims to unravel the importance of the partnerships design, structure, and leadership, as well as agency, connecting four types of conditions: (1) the features of the partnership, such as composition, governance structure, type, size, management, and leadership; (2) the agency-related and individual features of those involved, including the importance of trust, knowledge, expertise, and support; (3) the technological structures of the collaboration, especially ICT; and (4) the features related to the user's impact on the innovation process.

The following two parts of the book each consider a specific type of collaboration, showcasing in-depth empirical and original case-oriented analysis. Part II (Chapters 5–7) investigates intergovernmental collaboration for digital transformation, while Part III (Chapters 8–11) presents insights from several cases of public-private collaboration for digital transformation and innovation.

In Chapter 5, Breaugh and Nõmmik compare five case studies on collaborative management for government digitalization linked to the EU Single Digital Gateway legislation. The authors argue that system context to a large extent shapes the collaboration process and how both governance and leadership approaches unfold.

Chapter 6 by Rackwitz and van Doninck investigates two smart city initiatives in Belgium and Germany, both known as digital pioneers. The chapter explores the implementation and dynamic environments behind the collaboration in these initiatives, focusing on the who, the how, and the characteristics of collaboration and coordination in the two cases.

In Chapter 7, Rackwitz, Breaugh, and Hammerschmid give a systematic analysis of the leadership literature within public management and broader management, highlighting the most dominant theories related to leading collaborations. The chapter further draws on extensive qualitative data from ten case studies on collaborative digitalization projects in five countries and presents an in-depth study of key leadership strategies to handle collaborative complexity in intergovernmental collaborations.

In Chapter 8, Callens, Verhoest, Klijn, Brogaard, Lember, Pina, and van Doninck take on a practice-based view to analyse key success factors of 19 eHealth projects in five European countries. Through a qualitative analysis of more than 130 interviews with participants in these collaborations, the authors identify conditions that enable collaborative eHealth projects to succeed, and provide evidence of the salience of, for example, structural characteristics, agency-related features, the use of ICT, and user involvement.

Chapter 9 is about leading and managing complex innovation partnerships. Here, Callens and Klijn use a fuzzy-set QCA methodology to analyse 19 contract-based eHealth partnerships and investigate how contract conditions such as the presence of output specifications and contract flexibility work on the innovativeness of such services. The chapter also considers and tests the influence of network management strategies on the collaboration process on the innovativeness of these services.

In Chapter 10, García-Rayado and Callens analyse how public and private partners in eHealth innovation partnerships think that users should be involved. Using Q-methodology to analyse the perceptions of the partners, the authors look specifically at viewpoints concerning the motivations and activities of users, and viewpoints related to the support of the partnership for user involvement. Their findings show that multiple viewpoints exist, which has important implications for the design and leadership of such collaborations.

Chapter 11 by Langbroek and Verhoest looks more closely at the importance of interaction. By utilizing social network data from three different partnerships, it examines the extent to which the structure and interactions of collaborative partnerships in relation to the importance of actors can explain successful innovative outcomes.

The book ends with a reflective chapter, Chapter 12, by the editors, connecting the dots between the preceding conceptual, theoretical, and empirical chapters, and reflecting further on how we can use collaboration to enhance digital transformation.

NOTES

1. The TROPICO project received funding from the European Union's Horizon 2020 research and innovation programme under grant agreement No.726840. For more information: https://cordis.europa.eu/project/id/726840 (accessed 26 September 2023).
2. https://digital-strategy.ec.europa.eu/en/policies/europes-digital-decade (accessed 26 September 2023).
3. https://ec.europa.eu/info/strategy/priorities-2019-2024/europe-fit-digital-age_en (accessed 26 September 2023).

4. https://tropico-project.eu/case-studies/ (accessed 26 September 2023).

REFERENCES

Agger, A. and Sørensen, E. (2018). Managing collaborative innovation in public bureaucracies. *Planning Theory, 17*(1), 53–73.

Agranoff, R. (1986). *Intergovernmental management: Human services problem-solving in six metropolitan areas*. State University of New York Press.

Ansell, C. and Gash, A. (2008). Collaborative governance in theory and practice. *Journal of Public Administration Research and Theory, 18*(4), 543–71.

Batory, A. and Svensson S. (2018). *Literature and report review*, Deliverable 2.1, TROPICO project. https://cordis.europa.eu/project/id/726840/results (accessed 26 September 2023).

Bommert, B. (2010). Collaborative innovation in the public sector. *International Public Management Review, 11*(1), 15–33.

Breaugh, J., Rackwitz, M. and Hammerschmid, G. (2023). Leadership and institutional design in collaborative government digitalisation: Evidence from Belgium, Denmark, Estonia, Germany and the UK. *Government Information Quarterly, 40*(2). https://doi.org/10.1016/j.giq.2022.101788.

Callens, C. (2022). User involvement as a catalyst for collaborative public service innovation. *Journal of Public Administration Research and Theory 2022,* 1–13.

Callens C., Verhoest K. and Boon, J. (2022) Combined effects of procurement and collaboration on innovation in public-private-partnerships: A qualitative comparative analysis of 24 infrastructure projects, *Public Management Review, 24*(6), 860–81.

Camarinha-Matos, L.M., Fornasiero, R., Ramezani, J. and Ferrada, F. (2019). Collaborative networks: A pillar of digital transformation. *Applied Sciences, 9*(24), 5431.

Curtis, S. (2019). Digital transformation – the silver bullet to public service improvement? *Public Money and Management, 39*(5), 322–4.

Dockx, E., Verhoest, K., Langbroek, T. and Wynen, J. (2022). Bringing together unlikely innovators: Do connective and learning capacities impact collaboration for innovation and diversity of actors? *Public Management Review, 2022,* 1–24.

Dunleavy, P., Margetts, H., Bastow, S. and Tinkler J. (2006). New public management is dead: Long live digital era governance. *Journal of Public Administration Research and Theory, 16*(3), 467–94.

Emerson, K., Nabatchi, T. and Balogh, S. (2012). An integrative framework for collaborative governance. *Journal of Public Administration Research and Theory, 22*(1), 1–29.

European Commission. (2021). *Communication from the Commission to The European Parliament, The Council, The European Economic and Social Committee and the Committee of the Regions: 2030 Digital Compass: The European way for the Digital Decade*. European Commission. https://eur-lex.europa.eu/legal-content/en/TXT/?uri=CELEX%3A52021DC0118 (accessed 26 September 2023).

Gong, C. and Ribiere, V. (2021). Developing a unified definition of digital transformation. *Technovation,* 102(3). https://doi.org/10.1016/j.technovation.2020.102217.

Hanf, K. and Scharpf, F.W. (1978). *Interorganizational policy making: Limits to coordination and central control.* Sage Publications.

Hinings, B., Gegenhuber, T. and Greenwood, R. (2018). Digital innovation and transformation: An institutional perspective. *Information and Organization, 28*(1), 52–61.
Huxham, C. and Vangen, S. (2005). *Managing to collaborate – the theory and practice of collaborative advantage.* Routledge.
Kickert, W.J.M., Koppenjan, J.F.M. and Klijn, E.H. (1997). *Managing complex networks: Strategies for the public sector.* Sage Publications.
Klijn, E.H. and Koppenjan, J. (2016). *Governance networks in the public sector.* Routledge.
Lægreid, P. and Rykkja L.H. (2022). Accountability and inter-organizational collaboration within the state. *Public Management Review, 24*(5), 683–703.
Lægreid, P., Rykkja, L.H., Sarapuu, K. and Randma-Liiv, T. (2014). *Organizing for coordination in the public sector: Practices and lessons from 12 European vountries.* Palgrave Macmillan.
Lopes, A. and Farias, J. (2020). How can governance support collaborative innovation in the public sector? A systematic review of the literature. *International Review of Administrative Sciences, 88*(1), 114–30.
Mergel, I., Edelmann, N. and Haug, N. (2019). Defining digital transformation: Results from expert interviews. *Government Information Quarterly, 36*(4).
Molenveld, A., Verhoest, K., Voets, J. and Steen, T. (2020). Images of coordination: How implementing organizations perceive coordination arrangements. *Public Administration Review, 80*, 9–22.
Molenveld, A., Verhoest, K. and Wynen, J. (2021). Why public organizations contribute to cross-cutting policy programs: The role of structure, culture, and ministerial control. *Policy Sciences, 54*, 123–54.
Picazo-Vela, S., Gutiérrez-Martínez, I., Duhamel, F., Luna, D.E. and Luna Reyes, F. (2018). Value of inter-organizational collaboration in digital government projects. *Public Management Review, 20*(5), 691–708.
Pollitt, C. and Bouckaert, G. (2017). *Public management reform: A comparative analysis – into the age of austerity.* Oxford University Press.
Scott, I. (2020). Governing by silos. In *Oxford Research Encyclopaedia, Politics.* Oxford University Press. https://doi.org/10.1093/acrefore/9780190228637.013.1414.
Singh, A. and Hess, T. (2017). How chief digital officers promote the digital transformation of their companies. *MIS Quarterly Executive, 16*(1), 1–17.
Torfing, J. (2019). Collaborative innovation in the public sector: The argument. *Public Management Review, 21*(1), 1–11.
Vial, G. (2019). Understanding digital transformation: A review and a research agenda. *Journal of Strategic Information Systems, 28*(2), 118–44.

2. Digital transformation in the public sector

Miriam Lips

INTRODUCTION

Governments around the world are increasingly using digital technologies and data to fundamentally change their core functions, structures, operations, processes, activities, and relationships with stakeholders, including citizens, businesses, other government organizations, and civil society (Lips, 2020). These digital transformations are facilitated by fast-moving developments in the area of digital technologies, such as the Internet and other digital network infrastructures, social media, mobile technologies and devices, smart technologies, robotics, and artificial intelligence (AI). Another fast-moving technological development, which is further adding to the digital transformation potential available to governments and is particularly enabled by smart technologies and devices, is the creation of large volumes of data (e.g., Borgman, 2015; Gil-Garcia et al., 2016; Kitchin, 2014; Lips, 2020; Meijer and Bolivar, 2016; van Zoonen, 2016).

However, digital transformation in the public sector is not a straightforward process but can be seen as a complex issue. For instance, it is important to consider that digital transformation in the public sector is not caused by digital technology and data but is a complex process of mutual shaping between actors, technologies, institutions, economic factors, political factors, and socio-cultural factors (see also the next section). Also, digital transformation doesn't happen in a vacuum: in other words, governments haven't stopped their operations and activities in order to undertake a complete "reset" in the digital age, but they are adopting digital technologies and new forms of data use whilst they are dealing with business as usual (BAU).

Moreover, there are substantial risks involved for governments to undertake digital transformation initiatives, which are not attractive from a political leadership point of view: for example, many digital transformation initiatives in the public sector fail (Lips, 2020). Another substantial risk is that as the costs involved with digital transformation initiatives are so significant, many governments are reluctant to take on large-scale government technology projects that will have a major impact on their structures, functioning, and external relationships. Although the intended outcomes of digital transformation are difficult to achieve, it is clear that compared to the paper-based past, governments are fundamentally different in their operations as a result of digital transformation, which potentially could lead to increased effectiveness and efficiency. For example, a study conducted by McKinsey and Company (Dilmegani et al., 2014) suggests that capturing the full potential of digital transformation in the public sector could free up to US$1 trillion annually in economic value worldwide, through improved cost and operational performance.

In this chapter, we will further explore this concept of "digital transformation", and more specifically what it means in the public sector. First of all, we will explore a theoretical perspective on the process of digital transformation in the public sector, and in particular the role of technology, in the second section. In the third section, we look at how scholars have given meaning to the concept of digital transformation. We will be able to compare these scholarly meanings with forms of digital transformation in the public sector which have emerged in worldwide practice in the fourth section. And finally, not all digital transformation initiatives end up in failure: we can learn what has worked effectively from empirical research into successful digital transformation initiatives in the public sector. Ten success factors of digital transformation initiatives in the public sector will be discussed in the fifth section.

MUTUAL SHAPING OF DIGITAL TRANSFORMATION IN THE PUBLIC SECTOR

Digital transformation in the public sector commonly is seen as a development that started with governments' adoption of the Internet in the 1990s. However, the application of IT in government was not a new phenomenon at the time. Government computing had existed already for many years, with the first computers in government being used to perform simple mathematical calculations, such as data processing as part of the census in the USA (Agar, 2003; Lips, 2017). From the 1950s until the 1980s, governments started working with big, centralized mainframe computing machines for the execution of large-scale numerical processing tasks (Agar, 2003; Bellamy and Taylor, 1998; Lips, 2017). This particularly concerned government agencies responsible for storing and managing large quantities of data, such as tax authorities, social welfare departments, and national census bureaus. Introduced in government departments in the 1980s, personal computers (PCs) were initially used for word processing, calculation, data processing, and programming and became ubiquitous throughout government agencies (Bellamy and Taylor, 1998).

With the introduction of server-based networked computing, the PCs of public officials could be connected to the Internet, for example, which led to a shift away from centralized mainframe computing towards decentralized network computing in government (Lips, 2020). This trend continued with the introduction of more recent technologies. The Internet has already been mentioned as a digital technology that has an important impact on decentralized ways of working in government agencies, but also other digital network infrastructures, such as cloud computing, have had an effect of networked, decentralized access to data and information. Other more recent digital technologies that have a profound effect on government agencies are social media, mobile technologies and devices, smart technologies, robotics, and AI.

However, the influence of early IT solutions on government should not be underestimated. For example, governments continued to use centralized mainframe computing systems in parallel with decentralized network computing solutions after their adoption of the Internet. The operation and use of existing older digital technologies and IT systems in government is also referred to as "legacy systems". Legacy systems often have a profound impact on digital transformation initiatives, as governments have the costs of operating and maintaining these older operational systems whilst making investment decisions about the adoption of new technology.

However, although digital technology and data are prominent in digital transformation initiatives, they should not be seen as solely or in a linear way responsible for the societal outcomes: actual achievements in digital transformation initiatives in the public sector will not be determined by the technology per se, but by deep-seated social, institutional, legal, political, economic and cultural processes and structures, leading to fragmented and evolutionary outcomes (Bellamy and Taylor, 1998; Dutton, 1997; Fountain, 2001; Garnham, 1996; Lips, 2020). The relationship between digital technology and society can be characterized as a perspective of mutual shaping: digital technology and society influence and shape each other and, as a result, are mutually constituted rather than being neutral, mutually exclusive, and separate (Lips, 2020). Social choices between different technical options are made at every stage in the design, development, and use of the technology. These choices are not always conscious decisions; selected options depend on a wide range of technological and social factors, which, in interaction, co-shape the direction and outcomes of the technology itself and the societal change process it enables (Williams and Edge, 1996).

In general, when applied to digital transformation initiatives in the public sector, what this mutual shaping perspective reveals is a dynamic, complex, and incremental working together of digital technology capabilities, on the one hand, and socio-cultural, economic, political, institutional, and organizational factors, on the other (Lips, 2020). In other words, the design, development, and use of digital technology can be understood as a social phenomenon: an inextricable part of society (MacKenzie and Wajcman, 1999). As social factors are context-dependent and co-evolve with technology, societal change outcomes of this mutual shaping too are contingent and unique due to different social, cultural, institutional, and historical circumstances within a particular context (Lips, 2020).

This mutual shaping perspective, which is also known as the social shaping of technology (MacKenzie and Wajcman, 1985), explains why the same digital technology may be designed, developed, and used differently in different contexts at a particular moment in time, leading to varying outcomes (Lips, 2020). It also explains why there is often a mismatch between intentions for or expectations around a particular technology and the actual experience with this technology: in other words, why digital technologies often fail to deliver on predicted outcomes (Lips, 2020). In accordance with this mutual shaping perspective, the journey towards outcomes of a digital transformation initiative can be conceptualized as a complex negotiation process between various actors and technological, social, and institutional factors in a particular context (Latour, 1986). Several scholars also emphasize the critical role of users as co-shapers of digital technology within this mutual shaping eco-system (Dutton, 1999; Oudshoorn and Pinch, 2003; Williams and Edge, 1996).

This points to the critical importance of empirical observations and analysis in order to understand the mutual shaping of digital technology and government actors and their varying impacts, outcomes, and implications (Brown and Duguid, 2000; Castells, 1996; Woolgar, 2002). Empirical research in the context of digital transformation in the public sector has confirmed that digital technology- and data-related change in government should not be considered a rational, predictable, rapid, or straightforward process but an extremely challenging, unpredictable, complex, evolutionary, and non-linear course of action (Bellamy and Taylor, 1998; Eppel and Lips, 2016; Fountain, 2008; Lips, 2012; McLoughlin et al., 2013).

Digital technologies and data form critical parts of complex socio-technical assemblages positioned within the context of digital transformation initiatives in the public sector. Besides digital technologies and data, these socio-technical assemblages are made up of people and communities, users and stakeholders, government organizations and institutions, businesses, political leadership, public sector norms and values, democratic and ethical arrangements, legal and regulatory arrangements, budgets, policies, processes and procedures, forms of knowledge and experience and practices and activities (Lips, 2020). If we want to understand how digital technologies and data are being introduced, managed, and used in the context of digital transformation in the public sector, we will need to focus on the socio-technical assemblage as a whole and not just on the digital technology and/or the data per se (Ananny and Crawford, 2018; Bellamy and Taylor, 1998; Fountain, 2001; Kitchin, 2014; Lips, 2012, 2020).

Fountain offers an institutional perspective of the use of digital technology or enacted digital technologies as she calls it, in socio-technical assemblages in government and, with that, insights into the strategic role of digital technologies in the public sector in the following three ways (Fountain, 2001, pp. 195–6):

1. Enacted digital technologies are tools of politicians, public decision-makers, and public managers. These tools can be used or misused.
2. Enacted digital technologies become a vital digital infrastructure in governments. Once adopted, designed, and constructed, these digital technologies become part of the institutional settings of government organizations. Important influencing factors here are the high sunk costs of large digital transformation projects and the legislative arrangements that have been put in place to legally embed these digital technologies in the fabric of government. This situation not only influences future actions in government organizations but also creates a tendency to persist with this digital infrastructure even when technology changes rapidly, as institutionally embedded digital technologies are difficult to change.
3. Enacted digital technologies serve not only as enablers of change across government but even more so as strong catalysts for change within individual government organizations as they promise dramatic efficiency gains, productivity improvements, and more control. This strategic alignment between digital technologies and the interests of government organizations makes it difficult for public managers to resist using these technologies to produce such organizational promises; the institutional design of the bureaucratic state overwhelmingly structures behaviour within government agencies rather than between agencies.

Fountain (2001) also empirically observed at the time that in many cases in government, public officials used digital technologies in order to sustain or strengthen what she refers to as "deep institutions" or the history and culture encoded in the existing norms and values of a government organization. This phenomenon, which is also known as the "reinforcement thesis", acknowledges technology enactment by public sector decision-makers in ways that leave institutional structures and processes undisturbed or even reinforce and strengthen the institutional status quo (Fountain, 2001; Kraemer and King, 2006).

SCHOLARLY MEANINGS OF DIGITAL TRANSFORMATION IN THE PUBLIC SECTOR

In the last few decades, the transformational potential of applying digital technologies and data in government has strongly captured the interest of scholars (Taylor and Lips, 2008). Especially in the early days of the public Internet in the 1990s, but also more recently as a result of the arrival of new technologies, some form of radical transformation in the public sector was seen as an inevitable and often desirable outcome of the rapid technological developments that were happening in society at the same time (Lips, 2014; Taylor and Lips, 2008). A well-known example in the early days of the public Internet was that the application and use of the Internet would lead to a more horizontal, decentralized network society (Castells, 1996). However, as discussed earlier, the transformational potential embedded in digital technologies should not be considered a given; instead, digital transformation outcomes depend on the actual use of digital technologies and data as part of a particular socio-technical assemblage (Lips, 2020).

Moreover, although many scholars and practitioners have used "transformation" as a key concept to explore changes in the context of digital transformation initiatives, the meaning of this concept is usually ambiguous, ill-defined, or not defined at all (Lips, 2017; O'Neill, 2009). The meaning of the concept usually is associated with some form of fundamental change related to the application and use of digital technologies and data in the public sector (Lips, 2020). Based on empirical research in New Zealand, O'Neill (2009) offers a further distinction of the meaning of transformation in two separate ways:

1. *Instrumental transformation*: a radical change in the existing administration, information management, and service delivery practices of government agencies that may also have a consequential impact on organizational structures and/or management practices. This application of transformation often results in less disruptive changes to operational and management practices that deliver benefits of increased speed, better quality of government service, and lower transaction costs and can be described as "doing the same things differently".
2. *Systemic transformation*: a radical, disruptive change in existing governance arrangements in the public sector, including constitutional responsibilities and accountabilities, fiscal management, legislation, regulation, and decision-making rights over public resources. From this perspective, transformation is also about a disruptive, fundamental change in key institutional and democratic relationships, such as between government and citizens, within a broader systemic order. Therefore, systemic transformation is about "doing different things".

Several scholars have used the term "transformation" also as an ideal type for fundamental changes in the public sector related to the application and use of digital technologies and data: several maturity models for digital government have been distinguished, with the final stage being identified as "transformation". These maturity models all indicate a linear development towards better, mature, and transformed digital government (Andersen and Hendriksen, 2006; Klievink and Janssen, 2009). As an example, the following four-stage maturity model has been developed by Layne and Lee (2001):

- Stage 1 – *catalogue*: representing the presence of a government organization on the Internet;
- Stage 2 – *transaction*: users can do transactions online with the government;
- Stage 3 – *vertical integration*: the integration of a functional area across either one hierarchical government organization or different administrative levels; and
- Stage 4 – *horizontal integration*: the integration of processes and activities within and across different public sector organizations.

Another example is the following five-stage maturity model developed by Klievink and Janssen (2009, p. 278):

- Stage 1 – *Stovepipes*: few applications, public services or products are interconnected, and data are not shared between government organizations.
- Stage 2 – *Integrated government organizations*: public service delivery and digital technologies within government organizations are integrated to create a one-stop shop at the organizational level.
- Stage 3 – *Nationwide portal*: a nationwide government portal is introduced to provide access to existing government products and services, including a digital vault personalized for each individual citizen. The digital vault can be used to provide government organizations with access to citizens' personal data when their public services and products are being requested.
- Stage 4 – *Inter-organizational integration*: clearly defined and standardized cross-agency services are bundled and integrated and can be delivered as virtually one service via the portal.
- Stage 5 – *Demand-driven, joined-up government*: instead of citizens having to find and apply for government services, the portal will search for the relevant public services and make recommendations.

A further example of a maturity model that leads to digital transformation is the open government maturity model developed by Lee and Kwak (2012). This maturity model consists of the following five stages toward transformed open government (Lee and Kwak, 2012, p. 496–9):

- Stage 1 – *Initial conditions*: no or few open government capabilities exist, and social media are hardly used. The focus is on presenting and broadcasting information to the general public;
- Stage 2 – *Data transparency*: high-value, high-impact government data are published online and shared with members of the public. Data quality is assured in terms of accuracy, consistency, and timeliness;
- Stage 3 – *Open participation*: interactive communications with members of the general public are established around open government data, including participatory opportunities for public feedback, conversation, voting, and crowd-sourcing;
- Stage 4 – *Open collaboration*: inter-agency collaboration and open collaboration with members of the general public around open government data are established. The focus is on co-creating value-added government services; and
- Stage 5 – *Ubiquitous engagement*: public engagement around open government data becomes easier for members of the general public and universally accessible through mobile and ubiquitous computing devices and applications. Also, open government data, public engagement methods, social media tools, and government services are seamlessly integrated within and across government agencies so that the public can easily navigate and engage in various open government activities.

The benefits of achieving higher stages of digital transformation on these maturity models have been identified by Foley and Alfonso (2009). Based on research data from 28 digital government projects in one of the OECD member states, they found that the net benefits substantially increase when digital government projects are moving from the transaction stage into the transformation stage.

EMPIRICAL FORMS OF DIGITAL TRANSFORMATION IN THE PUBLIC SECTOR

From an empirical perspective, when we consider digital transformation initiatives where governments have fundamentally changed their core functions, structures, operations, processes, activities and/or relationships with stakeholders, including citizens, businesses, other government organizations, and civil society, we can observe the following forms of digital transformation around the world (Lips, 2020). First of all, forms of *service transformation* can be observed, where governments have fundamentally changed their public service provision to users. A particular form of service transformation is integrated government. This development is a shift away from government-centric ways of providing public services towards more customer-centric ways of public service delivery. The following three integrated government transformation areas can be distinguished:

1. Virtual integration, such as public services provided through integrated web portals;
2. Vertical integration of public services provided in different policy domains or government sectors, such as all public services provided in the wider social sector, through integrated government counters or "one-stop shops";
3. Horizontal integration of the front and/or back offices of government organizations.

Digital technologies, such as cloud computing, are considered critical enablers of joining up and integrating government front and back offices. A special mention needs to be made for digital transformation in the area of so-called shared services, or the horizontal integration of back-office services across government agencies in areas like finance, procurement, and human resources management.

Another form of service transformation is citizen-centric government. This form of service transformation became popular amongst governments in response to narrower customer-centric thinking in public service delivery. Citizen-centric government acknowledges the unique nature of the government's service relationships with citizens in the following ways (Lips, 2017; OECD, 2009a): firstly, the acknowledgement that citizens' rights and obligations are different from more narrowly defined customer rights; and secondly, that governments need to balance the distinct interests and needs of different groups of citizens within the broader framework of the public interest. This implies that governments cannot use a one-size-fits-all approach in their public service design and, consequently, need to shift from a traditional government-centric service provision paradigm towards a more differentiated citizen-centric service design paradigm (Lips, 2017; OECD, 2009a).

The main difference between a traditional government-centric paradigm and a citizen-centric paradigm is the following (OECD, 2009a): where a traditional government-centric paradigm takes an "inside-out" perspective by emphasizing the organizational logic and coherence of government, a citizen-centric paradigm takes an "outside-in" perspective by emphasizing the context of the citizen and their specific service needs. This then creates an external, citizen-centric logic for organizing government service provision and leads to the following characteristics of citizen-centric government (OECD, 2009a, 2009b):

- Taking an "outside-in" instead of an "inside-out" perspective;
- Individuals are not just customers of government agencies, but citizens with rights and duties;
- Many "customers" of government are "involuntary customers";
- An externally focused, citizen-centric logic for organizing government service provision;
- Providing integrated government services tailored to the individual needs of citizens;
- Allowing for cross-government data sharing to support personalized government service provision and remove duplicated efforts of citizens to provide personal data to government;
- Co-design of government services between government organizations and citizens;
- Collaboration between multiple service providers depending on the citizen's service needs; and
- A networked perspective of the wider public sector.

An example of how digital technologies can facilitate citizen-centric government is the application of the so-called life event model in public service provision. First introduced and used by the Singaporean government, this life event model clusters, integrates, and presents government information and services in accordance with major life events of citizens, such as looking for work, setting up a business, buying a house, having a child, and retiring (Lips, 2020). Another practical example of citizen-centric government is the introduction of administrative simplification initiatives in governments around the world, where citizens do not need any longer to submit the same personal data, such as the change of an address, over and over again to multiple government organizations but only need to tell government once (Lips, 2020).

Another form of service transformation is digital-by-default government. Especially popular during the recent Covid-19 pandemic, governments have replaced traditional paper-based and face-to-face service channels with a variety of digital channels (e.g., web services, smartphone apps, SMS, social media) and data as their standard way or default setting in government service provision (Lips, 2014). In general, there are six main reasons why governments want to become digital by default (Lips, 2014, p. 183):

1. To follow fundamental changes in society and seize the digital opportunity to transform themselves;
2. To promote substantial efficiency and cost savings;
3. To become customer-centric and provide better quality services;

4. To improve the availability and accessibility of open government data;
5. To move away from siloed approaches and duplications across the public sector by introducing common digital technology platforms; and
6. To promote the adoption of digital public services through enhanced security and privacy protection.

The following two main approaches can be observed in governments becoming digital by default (Lips, 2014): a service-centric approach, which converts traditional ways of government service provision into digital equivalents, and a data-centric approach, which makes open government data publicly available for use to create new information products, services, and applications that are most useful to consumers of that information. Another substantial difference in national digital-by-default government service reform strategies can be found in the implementation focus (Lips, 2020). For instance, some countries (e.g., the UK, Australia) focus on developing digital-by-default government services at the level of individual government organizations, whereas other countries (e.g., USA, Denmark) focus on developing government-wide digital-by-default solutions, such as open government datasets and public registers, which can be used in government-agnostic, programme-agnostic, and device-agnostic ways (Lips, 2014, p. 186).

Secondly, we can observe *business transformation* in government agencies around the world: the fundamental change in government organizations' structure, culture, processes, and activities. An important part of business transformation includes changes in how business is conducted, or the organization's operations, such as the change of traditionally paper-based systems into information systems which are enabling the operation of digital government. Good examples are the building of identity management systems, payment systems, and basic registrations or shared datasets across the government. Another example of business transformation is in the area of digital security.

SUCCESSFUL CONDITIONS FOR DIGITAL TRANSFORMATION IN THE PUBLIC SECTOR

Empirical research in New Zealand has pointed towards ten success factors for achieving digital transformation in the public sector. A successful digital transformation initiative can be identified as a project that is in use and has realized benefits for users and the project owner (Eppel and Lips, 2021a, 2021b; OAG, 2012).

First of all, having strong support from leaders and senior managers is an important success factor for digital transformation in the public sector. Ideally, strong effective leadership is both delivered at the political level and at the senior management level in government (Eppel and Lips, 2016; OAG, 2012;

Sikkut, 2022). Empirical research in six successful digital transformation projects in the New Zealand public sector showed that strong leadership and support from main stakeholders, such as ministers and senior managers (OAG, 2012, p. 53):

- are critical to effectively realizing benefits in digital transformation projects;
- can help to accelerate critical decisions, resolve resource blockages, set and manage realistic expectations, and add impetus to a project; and
- are particularly important for the reporting stages, when information about the project is given to people outside the project for them to make decisions – this includes the business case and benefits planning and reporting stages.

Effective leaders in digital government initiatives are often facilitative leaders, bringing people and stakeholders together and facilitating collaboration (Eppel and Lips, 2016, 2021a). Facilitative leadership was also required in the New Zealand-based Landonline case study as a result of the adoption of agile ways of working (Eppel and Lips, 2021a): it included senior executives finding ways to allow people working in development teams to take risks and make decisions within certain boundaries (guardrails) without having to first get permission from higher up the chain. The potential for a mistake or what was retrospectively seen as a poor decision was limited by the parameters of the guardrails and the ongoing stocktake and review processes of the agile methods used.

Based on an empirical study into effective leaders of digital transformation in government, Sikkut (2022, pp. 279–85) lists the following ten themes as lessons learned for delivering digital transformation best practice:

1. *Managing for delivery* – through remaking or creating policy, strategy, organizational structure, processes, team, culture, values, routines, effective leaders need to organize and lead for delivery. If effective leaders deliver, they will build credibility. This may be a prerequisite to be able to ask for levers or resources from superiors or build up soft power to ensure collaboration from other stakeholders, and get support from the public too (Sikkut, 2022, p. 280).
2. *Strategy setting* – a skilful strategist can use a strategy to build up the portfolio of activities and create a roadmap for lasting change. Setting a clear vision, an ambitious strategy, and an achievable action plan also can help to make delivery easier, because it can empower the team or rally other stakeholders behind (Sikkut, 2022, p. 281).
3. *Strong political support* – in some cases, strong political support from the highest-possible level in a country turned out to be the most fundamental precondition for successful digital transformation. Effective political

support goes beyond recognition, with political masters needing to take actual steps to give digital transformation leaders the institutional levers for coordination and delivery across the government or through the power of their office to mobilize stakeholders.

4. *Stakeholder management and networking* – to mobilize resources and to ensure necessary collaboration for delivery.
5. *Good communications* – effective communications can deliver benefits including strong public support, stakeholder buy-in, political backing, and talent recruitment. This was also an evident success factor in the Landonline and MyACC for Business case studies (Eppel and Lips, 2021a, 2021b).
6. *Building the best possible team* – to get the most potential out of a good team, people can be empowered with an enabling management style and routines, and a needed culture and values can be instilled.
7. *Working for lasting change* – lasting change is important as government agencies can be quite resistant to long-term change and fall back to previous ways of working. Most lasting change potential comes from building up strong and mature teams that can keep going without prior leaders, from leaving behind a strategy for the next years, and from building up a culture that is different from what was there before (Sikkut, 2022, p. 283). Another way of working on lasting change would be to embed changes into the machinery of government: policies, standards, laws, institutional arrangements, or international donor assistance deals are all ways to make it harder for successors to undo changes made. However, the best way to ensure continuity and make changes last is delivery: the creation of digital transformation solutions of a new quality that meet people's needs (Sikkut, 2022, p. 284).
8. *The job is hard: be ready for it* – if effective leaders want to keep delivering and achieve any impact, they should be sure to maintain their own stamina and keep a clear mind (Sikkut, 2022, p. 284).
9. *Fix your sight on the users: the impact on citizens and businesses* – effective leaders focus on the bigger picture: the mission and vision. This includes keeping their sight on the users and the impact they make on their lives through transformed digital services.
10. *Effectiveness is in the hands of the effective leader* – mindset and approach make all of the difference. Relevant experience, including an ability to handle challenges, is critical for this role.

Secondly, understanding the environment and making the most of the circumstances have been identified as an important success factor for digital transformation initiatives (OAG, 2012). More specifically, the empirical research found identifying increasing or future demand for services as an impetus for change and recognizing extreme or special circumstances of a given situation, including the impetus of limited time and treating this as an opportunity, as important ways to make the most of circumstances (OAG, 2012). The latter was, for instance, found in Christchurch with the establishment of the Christchurch Earthquake Support Syvstem, where digital services had to be delivered within a few days of the 22 February 2011 earthquake.

Thirdly, another success factor of digital transformation initiatives is using a business-led, flexible, and agile approach. The six successful projects that were studied in 2012 and the two more recent case studies published in 2021 all had a focused business purpose: there was an important acknowledgement in these projects that technology is not an end in itself but that a clear business purpose guides how the technology is used. In the case of the MyACC for Business case study, technology was seen as an important enabler of the envisioned organizational transformation roadmap, or a servant of business transformation. In this case, the use of technological solutions sat alongside organizational and cultural change and a central focus on the government agency's customers (Eppel and Lips, 2021b).

Because the six digital transformation projects and the two case studies all had a business focus, it was important to involve people who know a lot about the business, such as business owners. Other commonalities of the six successful projects were that they don't try to solve everything at the same time, they followed an iterative or pilot approach, and they used current technology where it makes sense to and did not reinvent solutions (OAG, 2012, p. 51). The two more recent successful digital transformation case studies also followed an agile, iterative approach. In the case of Landonline, during the transformation process, the organization had come to accept the transformational potential of the Landonline rebuild as an organization-wide challenge (Eppel and Lips, 2021a).

A fourth success factor of digital transformation initiatives is collaboration. Empirical research into six successful digital transformation projects shows that successful project teams have strong and collaborative relationships within the project team, with vendors, and with stakeholders, including other government agencies involved and end users (Eppel and Lips, 2016; OAG, 2012). The two more recent case studies point out the importance of internal collaboration between the business owners, the digital teams for achieving successful digital transformation and the teams responsible for customer engagement (Eppel and Lips, 2021a, 2021b).

Fifthly, a strong success factor of digital transformation initiatives is co-design and user-centricity. Empirical research into six successful digital transformation projects shows that successful project teams pay a lot of attention to users' experiences, and the two more recent case studies point out that the feedback and experience of users is critical to the development of the digital transformation solution, in line with the application of an agile approach (Eppel and Lips, 2021a, 2021b; OAG, 2012). Feedback from customers as the digital transformation solution was rolled out both allowed the product to be improved but also provided heart and encouragement to the developers that they are on the right track and building a product that is valued by customers (Eppel and Lips, 2021a).

Sixthly, of critical importance for successful digital transformation initiatives turned out to be the use of the right technology tools for a given set of circumstances or strategic, fit-for-purpose technology choices. The MyACC for Business case study also pointed to the requirement of a timely replacement of legacy systems. Empirical research into six successful digital transformation projects pointed at the following technology tools (OAG, 2012): agile methodologies, open data, and open-source technology tools. More recent empirical research confirms the critical importance of using agile methodologies in successful digital transformation initiatives (Mergel et al., 2018, p. 291): governments need to adapt to changes in their internal and external environments and create systems that allow them to scan trends, identify developments, predict their potential impact on the organization, and quickly learn how to implement changes to their standard operating procedures. Agile methodologies enable government agencies to become more flexible, adaptive, and rapid in their behaviour (Mergel et al., 2018). In the case of the Landonline case study, adopting an agile delivery method enabled an incremental approach to build fit-for-purpose in-house capacity and using user insights to inform the development of the system (Eppel and Lips, 2021a), multidisciplinary development squads were recruited to provide the capability to build quickly and well to meet user needs and work with end users on adoption. The MyACC for Business case study shows that the adoption of an agile approach was a deliberate decision in support of the agency's customer-centric strategy (Eppel and Lips, 2021b).

A seventh success factor in digital transformation initiatives is organizational openness to new ways of working. The decision to build the new Landonline product in-house necessitated some rethinking of how the responsible government agency organizes itself internally to do things and who is responsible for what (Eppel and Lips, 2021a). As the Landonline build became operational, it threw new light on business as usual and organizationally entrenched ways of working. In many cases, they were practices buried in (often long) history. A new technology application became an opportunity to consider whether they are still needed and the most efficacious.

An eighth success factor in digital transformation initiatives is fit-for-purpose governance and accountability. In the case of the Landonline case study, the initial governance process for the new system involved only a subset of the agency's senior executive team and some external digital governance experts who operated as a governance board (Eppel and Lips, 2021a). This arrangement risked overlap and conflicting responsibilities and accountabilities with the organization's executive leadership team. Gradual wider organizational realization of Landonline's transformational potential prompted a rethink about its governance: the original governance process was reformed to take account of Landonline's impact on the organization as a whole and included the full executive leadership team and the chief executive. The government agency also needed to change governance processes to fit with agile delivery methods, as was required in the MyACC for Business case study, and needed to convince Treasury and ministers of the need for different arrangements for accountability reporting, and funding drawdowns to match the more continuous delivery of an agile project (Eppel and Lips, 2021a, 2021b).

Ninthly, a success factor for digital transformation initiatives is change management and fit-for-purpose internal capability and processes. In the case of the Landonline case study, the government agency realized that it is not the technology alone that is changing but the organization is also changed by the technology and must be prepared to adapt and operate differently if full benefits are to be realized (Eppel and Lips, 2021a). A longer-term, capability-building view was taken in thinking about the capacity needed internally to build and maintain the new Landonline system.

Lastly, a tenth critical success factor in digital transformation initiatives is a clear articulation of the benefits, and routine monitoring of the benefits being realized (OAG, 2012). Effective benefits realization needs to be treated as a dynamic, continuous process of documentation and review of benefits. Using an agile methodology enabled the government agency to deliver benefits progressively along the way, rather than waiting some years for a fully developed product to be launched to realize those benefits (Eppel and Lips, 2021a, 2021b).

REFERENCES

Agar, J. (2003). *The Government Machine: A Revolutionary History of the Computer.* Massachusetts Institute of Technology Press.

Ananny, M. and Crawford, K. (2018). Seeing without knowing: Limitations of the transparency ideal and its application to algorithmic accountability. *New Media and Society 20*(3), 973–89.

Andersen, K.V. and Hendriksen, H.Z. (2006). E-government maturity models: Extension of the Layne and Lee model. *Government Information Quarterly 23*, 236–48.

Bellamy, C. and Taylor, J.A. (1998). *Governing in the Information Age.* Open University Press.

Borgman, C.L. (2015). *Big Data, Little Data, No Data: Scholarship in the Networked World.* Massachusetts Institute of Technology Press.

Brown, J.S. and Duguid, P. (2000). *The Social Life of Information.* Harvard Business School Press.

Castells, M. (1996). *The Rise of the Network Society: The Information Age: Economy,Society and Culture*, Vol. 1. Blackwell Publishers.

Dilmegani, C., Korkmaz, B. and Lundqvist, M. (December 2014). *Public-Sector Digitization: The Trillion-Dollar Challenge*. McKinsey and Company. Available at: https://www.mckinsey.com/business-functions/digital-mckinsey/our-insights/public-sector-digitization-the-trillion-dollar-challenge (accessed 2 November 2022).

Dutton, W.H. (1997). Multimedia visions and realities. In: Kubicek, H., Dutton, W.H. and Williams, R. (eds.) *The Social Shaping of Information Superhighways: European and American Roads to the Information Society*. Frankfurt: Campus Verlag, pp. 133–55.

Dutton, W.H. (ed.) (1999). *Society on the Line: Information Politics in the Digital Age*. Oxford University Press.

Eppel, E.A. and Lips, A.M.B. (2016). Unpacking the black box of successful ICT-enabled service transformation: How to join up the vertical, the horizontal and the technical. *Public Money and Management 36*(1), 39–46.

Eppel, E. and Lips, M. (2021a). *Digital Government Case Studies: LINZ Landonline. A successful digital government case study*. Victoria University of Wellington. Available at: https://www.wgtn.ac.nz/business/research/centres-institutes/chair-in-digital-government2/Digital_Govt_Landonline_Case_Study_Final.pdf (accessed 10 November 2022).

Eppel, E. and Lips, M. (2021b). *Digital Government Case Studies: My ACC for Business. A successful digital government case study*. Victoria University of Wellington. Available at: https://www.wgtn.ac.nz/__data/assets/pdf_file/0006/1938930/Digital-Government- Case-Study-ACC-Successful-Digital-Product-May2021.pdf (accessed 10 November 2022).

Foley, P. and Alfonso, X. (2009). eGovernment and the transformation agenda. *Public Administration 87*(2), 371–96.

Fountain, J.E. (2001). *Building the Virtual State: Information Technology and Institutional Change.* Brookings Institution Press.

Fountain, J.E. (2008). Bureaucratic reform and e-government in the United States: An institutional perspective. In: Chadwick, A. and Howard, P.N. (eds.) *The Routledge Handbook of Internet Politics.* Routledge, pp. 99–113.

Garnham, N. (1996). Constraints of multimedia convergence. In: Dutton, W.H. (ed.) *Information and Communication Technologies: Visions and Realities*. Oxford University Press, pp. 103–20.

Gil-Garcia, J.R., Pardo, T.A. and Nam, T. (eds.) (2016). *Smarter as the New Urban Agenda: A Comprehensive View of the 21st Century City*. Public Administration and Information Technology, Vol. 11. Springer International Publishing Switzerland. eBook, doi:10.1007/978-3-319-17620-8

Kitchin, R. (2014). *The Data Revolution: Big Data, Open Data, Data Infrastructures and Their Consequences.* Sage.

Klievink, B. and Janssen, M. (2009). Realizing joined-up government: Dynamic capabilities and stage models for transformation. *Government Information Quarterly 26*, 275–84.

Kraemer, K.L. and King, J.L. (2006). Information technology and administrative reform: Will government be different? *International Journal of Electronic Government Research 2*(1), 1–20.

Latour, B. (1986). *Science in Action: How to Follow Scientists and Engineers through Society.* Harvard University Press.
Layne, K. and Lee, J. (2001). Developing fully functional e-government: A four stage model. *Government Information Quarterly 18*(2), 122–36.
Lee, G. and Kwak, Y.H. (2012). An open government maturity model for social media-based public engagement. *Government Information Quarterly 29*, 492–503.
Lips, A.M.B. (2012). E-government is dead: Long live public administration 2.0. *Information Polity 17*(3–4), 239–50.
Lips, A.M.B. (2014). Transforming government: By default? In: Graham, M. and Dutton, W.H. (eds.) *Society and the Internet: How Networks of Information and Communication Are Changing Our Lives.* Oxford University Press, pp. 179–94.
Lips, A.M.B. (2017). Transforming government services over time: Meanings, impacts, and implications for citizen-government relationships. In: Chen, Y. and Ahn, M.J. (eds.) *Routledge Handbook on Information Technology in Government.* Routledge, pp. 11–26.
Lips, M. (2020). *Digital Government: Managing Public Sector Reform in the Digital Era.* Routledge.
MacKenzie, D.A. and Wajcman, J. (1985). *The Social Shaping of Technology: How the Refrigerator Got Its Hum.* Open University Press.
MacKenzie, D.A. and Wajcman, J. (eds.) (1999). *The Social Shaping of Technology*, 2nd edn. Open University Press.
McLoughlin, I. and Wilson, R., with Martin, M. (2013). *Digital Government @ Work: A Social Informatics Perspective*. Oxford University Press.
Meijer, A.J. and Bolivar, M.P.R. (2016). Governing the smart city: A review of the literature on smart urban governance. *International Review of Administrative Sciences 82*(2), 392–408.
Mergel, I., Gong, Y. and Bertot, J. (2018). Agile government: Systematic literature review and future research. *Government Information Quarterly 35*(2), 291–8.
O'Neill, R.R. (2009). *E-Government: Transformation of Public Governance in New Zealand?* PhD Thesis.Victoria University of Wellington. Available at: http://researcharchive.vuw.ac.nz/handle/10063/929 (accessed 4 November 2022).
OECD (2009a). *Citizen-Centric Government.* OECD.
OECD (2009b). *Rethinking e-Government Services: User-Centred Approaches*. OECD.
Office of the Auditor-General (OAG) (2012). Realising benefits from six public sector technology projects. Wellington: OAG.
Oudshoorn, N. and Pinch, T. (eds.) (2003). *How Users Matter: The Co-construction of Users and Technology.* Massachusetts Institute of Technology Press.
Sikkut, S. (2022). *Digital Government Excellence: Lessons from Effective Digital Leaders.* John Wiley and Sons.
Taylor, J.A. and Lips, A.M.B. (2008). The citizen in the information polity: Exposing the limits of the e-government paradigm. *Information Polity 13*(3–4), 139–52.
Williams, R. and Edge, D. (1996). The social shaping of technology. In: Dutton, W.H. (ed.) *Information and Communication Technologies: Visions and Realities.* Oxford University Press, pp. 53–67.
Woolgar, S. (ed.) (2002). *Virtual Society? Technology, Cyberbole, Reality.* Oxford University Press.
van Zoonen, L. (2016). Privacy concerns in smart cities. *Government Information Quarterly 33*, 472–80.

3. Intergovernmental collaboration in the context of digital transformation: state-of-the-art and theoretical notions

Gerhard Hammerschmid, Jessica Breaugh and Maike Rackwitz

INTRODUCTION

The process of digitalisation is rapidly changing the way public sector organisations are managed and governed. In order to achieve the increasing pressures towards digital government, more and more governments are turning towards collaborative approaches, not only because many digitalisation processes cross ministerial lines, but also to harness the interest, knowledge, and skills of their own public service. Simply put, digital transformation challenges silo mentality, forcing government organisations to work together to build products to service citizens. Digitalisation processes also necessitate a complete re-evaluation of the way government policies, processes, and services are designed and implemented (Mergel et al., 2019). However, notwithstanding the need for collaboration, being able to manage large-scale collaboration projects in the realm of digitalisation is notoriously difficult and requires the ability to manage both the technologies being implemented, as well as the people who are expected to implement them. Indeed, there is always a balancing act between the push of technological change and the pull of organisational capacity (Clausen et al., 2020).

The process of digitalisation, in conjunction with intergovernmental collaborative approaches, does possess characteristics that are rather unique. These include the technical nature of the projects, where success hinges on both the technical capabilities as well as managing the many stakeholders involved in the project (Picazo-Vela et al., 2018). The combination of outside pressures for more digital and streamlined services and the speed at which technology is changing offers a double burden on governments – racing to find solutions while also understanding the need to work through unclear or archaic regulatory processes, legal and constitutional challenges as well as traditions of ministerial independence. Supra-national reform pressures, for example, from the EU, add additional pressures and complexities when it comes to designing interoperable citizen services that are accessible to all EU residents.

Despite the difficulties of collaborative projects, the collaborative approach has become an essential component of modern public organisations, as many move towards 'collaboration by default' (Ansell and Gash, 2008; Dickinson and Sullivan, 2014; Torfing, 2019). Given the complexity that digitalisation processes entail, including handling the complexity of large-scale and multi-actor collaboration projects, it is necessary to better understand how this can be achieved. This includes both understanding the system challenges and related dynamics of collaboration such as complexity, risk, and power, as well as the public management tools that can be used to manage these challenges and dynamics such as institutional design and leadership. The purpose of this chapter is to do just this – examine how public management interventions can be used to manage the systemic challenges of intergovernmental collaboration. To do so, we start with a review of the collaboration literature, focusing on collaborative management, collaborative innovation, and digital innovation. We then shift to exploring the role of system context as a key starting condition in understanding collaborative dynamics. Next, we examine key challenges in collaboration projects as well as key public management interventions related to institutional design and leadership that can be used to tackle these challenges. Before outlining how this framework can be applied in varying contexts, we also note the difficulties in measuring performance outcomes.

INTERGOVERNMENTAL COLLABORATION AND DIGITAL GOVERNANCE

The research field in government collaboration is a fast-moving and disjointed field of research (Bianchi et al., 2021) primarily due to the variety of different foci and terminology used to understand and study collaborations (Scott and Merton, 2021). In an effort to clarify our research scope, we briefly review three areas of collaboration research. The first is collaborative management or the structures and design features of collaborative endeavours, the second is collaborative innovation, focusing more specifically on the role that collaboration plays in handling wicked public management problems, and finally, digital governance, which focuses especially on how digital projects have been adopted and governed in the public sector.

COLLABORATIVE MANAGEMENT

The concept of collaborative management has become highly salient within the context of digitalisation primarily due to its ability to understand and explain cross-boundary transformative projects by examining the structures in which they are embedded (Agranoff and McGuire, 2003; McGuire and Agranoff, 2011). The empirical research focusing on collaborative management has, for the most part, substantiated the theoretical propositions from the literature. In most cases, empirical scholars emphasise tangible project-related characteristics such as institutional design and leadership as critical components in collaborative management (Eriksson et al., 2020) in addition to trust and relationship building (Cristofoli et al., 2017; Vangen and Huxham, 2003). More and more scholars are now coming to the conclusion that there are many critical elements in collaborative management approaches. These include how interactions are structured. Clear rules and mandates, goal commitment, reducing transaction costs as well as being able to manage interpersonal relationships appear to be critical (Eriksson et al., 2020; Krogh, 2022; Scott and Merton, 2021).

COLLABORATIVE INNOVATION

In contrast to collaborative management, collaborative innovation is a term emerging from the innovation literature, with a focus on using innovative solutions to handle difficult, hard-to-define, and hard-to-solve problems facing society (Torfing, 2019). Often referred to as 'wicked' problems, built upon the work of Rittel and Webber (1973), this type of collaborative innovation focuses on the relationship between various governmental and non-governmental stakeholders and the decision-making processes undertaken to address complex challenges facing society (Wegrich, 2019, p. 12). Similar to the literature on collaborative management, there is a strong focus on stakeholder management and critical leadership skills, but also introducing elements of risk-taking (Bommert, 2010; Krogh and Torfing, 2015; Torfing et al., 2019).

DIGITAL GOVERNANCE

Collaborative management focuses on how management structures can be used to handle the complexity related to working across government silos, while collaborative innovation takes this one step further in addressing how these collaborations can be used to harness innovative programme design and delivery for increasingly complex problems. While the digitalisation process is indeed a complex endeavour involving a multiplicity of actors, for the most part, they are more 'tangled' than 'wicked' (Dawes et al., 2009). Yet, the rapid and often disruptive changes occurring within the context of digitalisation present themselves not only as a shift in the way services and programmes are delivered from analogue to digital. They also manifest in a de-siloing process inside government (Margetts and Dunleavy, 2013, p. 6), needs-based holism, or focusing on design elements for citizens rather than government to increase their successful adoption (Mergel et al., 2019). The complexity of governance then becomes designing and managing internal and external changes happening simultaneously. It is a process that requires many different types of skills, and also, without an end (Mergel et al., 2019; Picazo-Vela et al., 2018), challenging traditional project management logics (Clausen et al., 2020). The key question then becomes, how can these processes be understood and governed? The next section sets out a preliminary framework to do this.

CONCEPTUAL FRAMEWORK

Based on the established literature on collaboration and digital governance discussed above, several common themes emerge related to collaboration dynamics.

System Context

While collaborative approaches to digitalisation offer a means of addressing essential whole-of-government problems, the context in which the collaboration takes place also plays an essential role in the challenges, dynamics, and solutions that emerge as well as the ultimate success (or failure) (Emerson et al., 2012). Indeed, the system context not only impacts the entire collaborative process but many of the barriers that emerge typically can be identified from the system context itself – especially in the start-up phase of the collaboration (Mergel, 2018). Elements of system context include the political system in which the collaboration is embedded, the political and administrative context (including political ripeness), as well as legal foundations of the state, the organisational culture and logics of specific stakeholders, and the stakes of the overall project (Ansell and Gash, 2008; Bingham, 2008; Dawes and Pardo, 2002; Emerson et al., 2012; Jaeger, 2002; Lewis et al., 2018). These types of contextual factors impact the challenges and dynamics that emerge within a collaboration project.

Challenges and Dynamics

In all forms of collaborative endeavours, challenges and dynamics impact the way the collaborations are developed and managed. In the public sector, in particular, collaborative networks challenge traditional silos of government, including introducing overlapping authorities and difficulties in performance measurement and accountability (Bovens, 2007; Klijn and Koppenjan, 2014). In a recent literature review, three key aspects that challenge collaborations, in particular, emerged (Rackwitz et al., 2020). These are complexity (Klijn and Koppenjan, 2014), risk (Brown and Osborne, 2013; Timeus, 2019), and power (Ansell and Gash, 2008; Huxham and Vangen, 2015).

Complexity

Complexity refers to complications arising from managing different hierarchical and/or legal structures, relationship management, and changing natures of project goals and perceptions (Klijn and Koppenjan, 2016; Meijer, 2015). Following Klijn and Koppenjan (2016), complexity can be divided into three sub-types: substantive, strategic, and institutional. *Substantive complexity* refers to the variation in perceptions of the collaboration's goals, problematisation, and solutions by project partners. This type of complexity can lead to a lack of shared meaning, leading to coordination problems related to selective perceptions, or those whose goals and associated behaviours are pursued individually rather than as a collective (Dearborn and Simon, 1958; Klijn and Koppenjan, 2014; Wegrich, 2019). *Strategic complexity* focuses on the variance in strategic direction within a collaboration. *Institutional complexity* refers to the institutional conditions that inherently guide collaborative frameworks and constrain behaviours. Typically, this reflects different legal or policy precedents, and also formal organisations' and/or institutions' roles and regulations. Within the context of digitalisation, issues such as designing appropriate technologies (Liu and Zheng, 2018; Mergel, 2018), a lack of IT skills and technological dependencies and information asymmetries add to these complexities (Dawes and Pardo, 2002; Ferro and Sorrentino, 2010; Neumann et al., 2019) and lead to both strategic and substantive complexity. Institutional complexity is also particularly acute due to the cross-cutting nature of the government digitalisation project, especially when they cross or challenge traditional legal or jurisdictional boundaries (Apostolou et al., 2011; Mergel, 2018).

Risk

Within the context of collaboration, risk is a second key challenge, or rather, the perception of risk (Timeus, 2018). Digital projects entail a mix of open and closed system risks depending on the level of analysis. For certain project partners, the internal risks can be minimised through processes and procedures, while at the project level, a certain level of risk is accepted and managed in order for the innovation to be brought ahead, which also means exploiting the innovation process in and of itself (Brown and Osborne, 2013; Hartley et al., 2013). What does set digitalisation projects apart are their high risks of failure, working in unregulated territories and high financial investments (Mergel,

2016; Neumann et al., 2019). The failure to actively manage risk may also increase risk perceptions – especially if and when collaborative projects have a diversity of actors and their values, goals, and cultures clash (Gil-García et al., 2019). This is supported by Wang and Feeney (2016), who note that a risk-taking culture needs to be established to mitigate project-related challenges, structural constraints, and the adoption of technology.

Power

The concept of power relations is critical for understanding the behavioural dynamics of actors engaged in collaborative endeavours. What is critical is not necessarily power itself, but whether or not there is a negative perception of power imbalance within a collaborative relationship. The perception of negative power imbalances can hamper the collaborative process and also impact the way goals are achieved and resources are allocated (Ansell and Gash, 2008; Choi and Robertson, 2014; McGuire and Agranoff, 2011). Because of this, managing power imbalance becomes a critical factor upon which success hinges (Huxham and Vangen, 2005; Klijn and Koppenjan, 2000; Purdy, 2012; Ran and Qi, 2018, 2019; Sedgwick, 2016). Such power imbalances can operate in two ways. First, if power imbalances are the result of a lack of capacity or knowledge or second, if they are due to unequal power distributions mandated in the projects themselves (like coordinating bodies) (Ansell and Gash, 2008; McGuire and Agranoff, 2011). However, these can be managed through both structure and governance mechanisms that clearly outline power distributions and means of correcting and controlling this power, making power more complementary. Research examining power in digitalisation projects is inconclusive. Some have found that dispersed power relations made collaboration the only solution to achieve project-related goals (Luna-Reyes et al., 2007), while others showed that power imbalances hampered the diversity and inclusion necessary for the intended collaborative process (Jones and Hooper, 2017).

PUBLIC MANAGEMENT INTERVENTIONS

In order to manage complexities of large-scale intergovernmental collaborations, scholars have pointed to institutional design features and leadership as essential (e.g., Ansell and Gash, 2008; Torfing, 2019).

Institutional Design

With respect to institutional design, we focus on structures, level of inclusion, process legitimacy, and trust building (Torfing, 2019; Scholl et al., 2012). Considering that the adoption of digital processes in government is often enacted and constrained by structure (Fountain, 2001), it is important to outline different types of structures. Hierarchical structures reflect traditional, top-down structures. Design features of this approach include strict accountability and governance through legal regulations. Collaboration challenges are thus handled through formal institutional rules and processes (such as a strict chain of command, planning, compliance and control), concentrating power at the top (Meuleman, 2008). The utility of this style is mixed, with some noting a certain level of malaise and constraint to collaborative capacity (Kickert et al., 1997; Wegrich, 2019) while others have noted the necessity of top-down control as a function of government (Span et al., 2012). Inspired by New Public Management, market-based structures use market-style principles such as a focus on cost savings, input-output calibrations, professional management, and measures of performance (Hood, 1991). This approach manages collaboration complexities via the heavy use of negotiation tactics and decentralised power distributions. The final structure that we focus on is network governance structures. These are characterised by flat structures, a sense of cooperation, equality, and a lack of formalised leadership (Provan and Kenis, 2008). Collaborative complexity is handled through inclusion and a sense of community building forming a shared participant model making power shared and symmetrical across all stakeholders (Meuleman, 2008). While these structures are presented with a certain level of mutual exclusivity, in reality, empirical research has shown that hybrid structures may also emerge at different stages of the projects (i.e., more structure at the beginning of the project, moving towards a more networked-based model later on). However, many still advocate for a lead organisation model, meaning one organisation takes on a clear management and coordinating role, to ensure smooth coordination, especially if projects are large and cross-cutting (Clarke, 2019; McGuire, 2006; Span et al., 2012).

Inclusion

A second component of institutional design critical for intergovernmental collaboration is the concept of inclusion. According to scholars, widespread inclusion is a key component of collaboration projects to build legitimacy into the process, enhance creative and innovative potential, and ensure that agreed-upon goals and project outcomes take into consideration all stakeholders in the process (Ansell and Gash, 2008). Establishing who the stakeholders are from a very early stage also helps to identify roles and responsibilities towards the strengths of these stakeholders, and also reduces ambiguity in who is involved and who is not (Vangen and Huxham, 2003).

Rules

A third component is clear protocols and rules that guide behaviours of actors within the collaboration (Ansell and Gash, 2008). Indeed, rules provide a clear purpose, and rules and expectation management are critical components of collaborative digitalisation projects (Dawes and Pardo, 2002; Gil-García et al., 2018). These ground rules serve several purposes. First, they help to guide behaviour, by providing information to actors about what is acceptable and what is expected behaviours, which can reduce both scepticism, risk, and negative perceptions of power imbalances (Ansell and Gash, 2008). Second, clear rules identify boundary conditions that clarify the roles and responsibilities of stakeholders including the intended outcomes of the collaborations themselves (Vangen and Huxham, 2003). This leads to better goal alignment which then reduces complexity and risk from participating actors (Scholl et al., 2012), and can foster a greater sense of goal commitment which may lead to higher levels of collaborative persistence (Scott and Merton, 2021). Clear rules are also an essential foundation for trust to develop, where any potential problem and/or power struggle can already be integrated into the rules themselves (Wegrich, 2019). A lack of ground rules can, and has, led to planning difficulties (Gil-García et al., 2007).

Trust

Building and maintaining trust is a final component of handling complexity related to collaboration, primarily due to its link with performance outcomes (Chen et al., 2019). Understood as a type of 'soft factor', it focuses on relationship building and building norms of reciprocity (Cristofoli et al., 2017; Emerson et al., 2012; Eriksson et al., 2020) which acts as a 'lubricant that enhances mutuality' (Sedgwick, 2016, p. 238). According to Vangen and Huxham (2003), collaborative relationships involve trust, willingness to collaborate, and actual work being completed. They are all interdependent on one another and are built through a 'cyclical trust-building loop' (McGuire, 2006, p. 38). Building trust over time is essential for overcoming barriers in the collaborative process, and also because it is a way to signal actors' intentions (Klijn and Edelenbos, 2007). Initial trust-building exercises always come with an element of risk, however, and begin to build as expected and actual behaviours align (Vangen and Huxham, 2003). In some instances, trust is a means of reducing risk perceptions (Cohen and Cohen, 2021). This also increases the willingness to take active roles in collaboration projects (Luna-Reyes et al., 2007). Because trust is linked to social capital development, it is also linked to stronger project norms and behaviours – facilitating decision-making processes (Fedorowicz et al., 2014).

The collaborative literature focusing on digitalisation also emphasises trust building as a core component of managing collaborations and notes that inter-agency trust is directly related to performance outcomes (Chen et al., 2019). Some emphasise that trust is important for building a sense of shared meaning and its effects are reflected in all aspects of the collaborative project, including being able to work together and correct problems (Liu and Zheng, 2018). Luna-Reyes et al. (2007) also note that trust is one of the most important components in collaborative arrangements of digitalisation projects. They argue that this is affected by institutions themselves and that trust results in the interest and willingness of actors to collaborate. They especially provide evidence that showing results early in the process helps to build trust. In addition, institutional arrangements, organisational structures, and management processes reinforce trust and decrease the perceptions of risk.

Dawes and Pardo (2002) state that relationship building does not just occur at the onset of a new collaborative digital project, it can be based on previous collaborative experiences, and also that these relationships may dictate the current ones (i.e., if they were collaborative, contractual, regulatory, or adversary). As a result of this, history is an important consideration. They argue that to build trust in collaborative digital systems, individuals and organisations need to be able to identify and understand stakeholders, understand their 'abilities, strengths and limitations' (p. 269), and commit to partnerships with some level of interdependence and division of tasks and responsibilities. Chen and Lee (2018) take this distinction one step further to identify trust relationships at different levels of the collaboration itself. To gain trust and support, financial support can help get people on board with the programme and build institutional arrangements for joint action. As social capital increases, so does the strength of project norms and behaviours, which results in a higher probability of coming to agreements on technological and process decisions (Fedorowicz et al., 2014). Soft factors involving trust and interpersonal relationships are a recurring theme in both public management and digital governance literature as it is a design feature that can be used to handle collaboration challenges related to complexity, risk, and power imbalances in the context of digitalisation efforts.

LEADERSHIP

In addition to institutional design mechanisms, the role of leaders in steering large-scale collaboration projects is paramount (Lewis et al., 2018; O'Leary et al., 2012). While there is a plethora of leadership literature, a specific subset of this literature especially looks at leadership in the context of collaboration. Collaborative leadership scholars have noted that active participation, control and facilitation, and group management skills are critical components of being a successful leader (Lasker et al., 2003). As a result, leaders are often tasked with managing both overarching project complexities and also handling interpersonal dynamics on top of this (Torfing, 2019).

In the literature, there are different types of leadership approaches that characterise collaborative leadership. These include engaging in the role of convenors, facilitators, mediators, and catalysts (Ansell and Gash, 2012; Torfing, 2019). Convenors play the role of facilitators, ensuring inclusivity, empowerment, and active information exchange (Torfing, 2019). Facilitators play three different roles as stewards, mediators, and catalysts. Stewards focus on the management components of leadership – that is, setting ground rules, developing a sense of shared goals, and encouraging the development of social capital (Ansell and Gash, 2012). Mediators have a stronger focus on interpersonal relationship building and have skills in brokering and conflict management by clarifying goals, understanding different perspectives, and understanding and managing power dynamics (Ansell and Gash, 2012; Crosby and Bryson, 2010). Catalyst leadership approaches are more active than mediators by playing an active role and being a benefactor of problem-solving solutions. Because of this they seize opportunities and are able to see and understand the larger role that each stakeholder plays (Ansell and Gash, 2012; Morse, 2010).

Despite the beneficial aspects of collaborative leadership approaches, public organisations also benefit from transactional types of leadership, due to the fact that public organisations remain hierarchical in nature. This hierarchy, however, can be used as a means of problem-solving (Meuleman, 2008). Transactional leadership characteristics focus on procedurally oriented behaviours and look at how incentive structures can be used to motivate individuals (Jensen et al., 2019; Lewis et al., 2018). Linked to this are clear goals and behavioural expectations (Ruggieri and Abbate, 2013, p. 1172).

To reconcile both collaborative and transactional leadership styles, many scholars have now begun to question the either-or rhetoric, and instead stress the importance of a multi-faceted or contingent leadership approach in managing collaborative projects (Ansell and Gash, 2012; Eriksson et al., 2020; Van Wart, 2003). This means they need to be able to adapt to stress, change, and problems that may emerge from the collaborative process. They essentially become critical for both nurturing the development of the collaboration itself (using, e.g., collaborative leadership skills), but also buffering instability or conflict diffusion (through, e.g., engaging in more transactional leadership skills) (Agger and Sørensen, 2018; Vangen and Huxham, 2003). This is very similar to Clarke (2019) who makes reference to a top-down versus diffused leadership model with top-down referring to standard-setting and authoring building, whereas diffuse leadership focuses more on relationship building through support and guidance. A common thread in this, however, is the reactive ability of a leader, especially in seeing gaps, understanding the interplay and incentives of actors, the roles and responsibilities of partners, as well as the overarching 'picture' of a large-scale collaborative project to be able to manage across networks (Chen and Lee, 2018; Elnaghi et al., 2019; Jones and Hooper, 2017; Scholl et al., 2012). In some cases, the leader's ability to shift and adapt design features to reflect changing contexts is also critical (Krogh, 2022). For more information regarding leadership, see Chapter 7.

OUTCOMES

Measuring performance outcomes is a notoriously difficult undertaking, primarily due to public sector logics that favour public value creation over competition (Moore, 1995). Within the context of collaboration, projects tend to be rather long, with limited goal clarity further complicating the ability to measure direct outcomes (Clarke, 2019; Scholl et al., 2012). Yet outcomes are essential components for collaborations, not only as a means of tracking the impact but also because they play an important role in future collaborations in terms of the ability to develop good working relationships (Ansell and Gash, 2008; see also Chapter 5). Despite this rather underreached or underreported area of collaboration (de Vries et al., 2016; Scholl et al., 2012; Torfing, 2019), some scholars offer a framework for study outcomes, nonetheless. More specifically, Ran and Qi (2018) outline two approaches to measure collaboration outcomes. The first is project-related goals (or outcome-based measures), while the second focuses on the collaborative processes that enabled the project to achieve its goals (process-based measures) (Head, 2008). Process-based measures are sometimes linked to the notion of 'small wins' that positively reinforces collaborative behaviour and persistence (Doberstein, 2016; Termeer and Dewulf, 2019), or as Mergel et al. (2018, p. 295) note, early wins can also be linked to 'early validation' reducing the risk perceptions of collaborative actors, and also initial evidence of a viable digital product.

Outcomes are important to examine as they set a precedent for future projects, build confidence in the collaborative potential, and increase the appeal of the collaborative approach (Luna-Reyes et al., 2007; Scholl et al., 2012). It is therefore essential to include outcomes in the conceptual framework.

OVERARCHING FRAMEWORK

Based on the development of the previous sections, we reiterate the conceptual model elaborated by Rackwitz et al. (2020) (Figure 3.1). First, complexity, power, and risk are central to the model, highlighting their critical and often oscillating influence on collaborative outcomes. Figure 3.1 also underscores the fact that these three aspects of collaboration are not mutually exclusive from one another – rather, they often influence each other. The model also shows the variety of systemic contextual factors that influence or are influenced by collaboration challenges and dynamics. As per our theoretical stipulations, it also shows the moderating role that public management interventions such as key institutional design features as well as leadership has on collaborative outcomes. This means we theorise that particular types of public management interventions have the potential to enhance or thwart the collaborative challenges and dynamics of intergovernmental collaboration.

By presenting it in this way, we open up the possibility of testable hypotheses for studying elements of intergovernmental digital collaboration projects, including studying specifically which types of context factors impact complexity, power (im)balance and risk perceptions, and which design features or leadership skills are more suited to handle this. Chapters 5, 6, and 7 begin to examine some of these relationships in more detail.

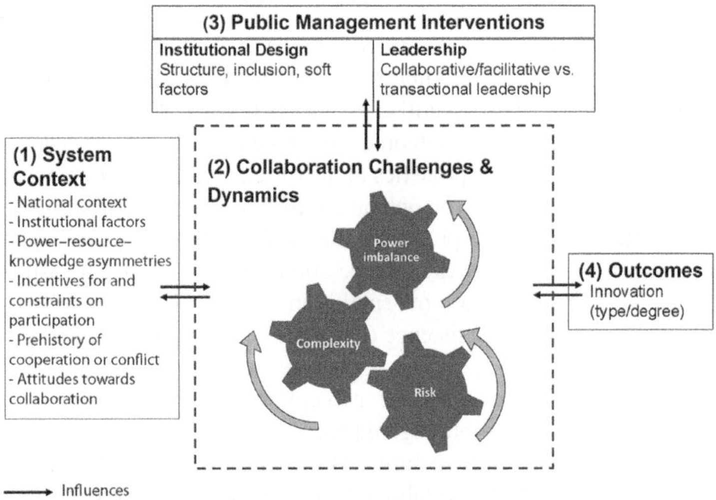

Source: Taken from Rackwitz et al. (2020) based on research by Ansell and Gash (2008); Brown and Osborne (2013); de Vries et al. (2016); Emerson et al. (2012); Hartley (2005); Klijn and Koppenjan (2014); Ricard et al. (2017); Torfing (2019).

Figure 3.1 Conceptual model for collaboration dynamics

APPLICATION AND TESTING OF THE FRAMEWORK

The application of our proposed model requires a variety of research methods and modes of analysis, especially with respect to the more nuanced aspects of each of its components. This is particularly relevant to testing its applicability and universality. It also opens up the possibility to examine digital projects and their development at the process level (Mergel et al., 2019). This then enables one to examine aspects related to interpersonal behaviours including perceptions and behavioural decision making (which plays a very important role in each component of our conceptual model).

The first way in which the model can be tested is by examining the interplay between complexity, power, and risk. This can be accomplished in several ways, including examining the level of relevance and intensity of these challenges in a particular case and then validating this variance through a multi-case study design. For example, one could select cases with different levels of challenge intensities, and then analyse if public management interventions also change. This may reveal not only how particular behaviours emerge, such as turf wars, but also, how they are managed. A multi-case study design could then test the external validity and reliability of our claims, something that is already rather sparse in the digitalisation literature (Neumann et al., 2019). This may also provide systematic and empirical evidence to examine the extent to which collaboration challenges and dynamics are universal.

Linked to the notion of universality, a cross-country study could also exploit the rather rich literature on the role of administrative traditions in governance approaches (Guy Peters and Pierre, 2019). Within the European context, these include, for example, Anglo-Saxon, Continental, and Scandinavian traditions (Benz and Zimmerman, 2011; Huxley et al., 2016; Meyer and Hammerschmid, 2010). Traditions are an interesting way to analyse collaboration arrangements because of their ability to explain the variation in collaborative behaviours by using institutional logics. This includes, for example, the legal frameworks of a country, an understanding of hierarchy and the level of pragmatism used in governance. These are the same characteristics which we argue are critical for understanding the starting conditions of intergovernmental collaboration because they dictate, to a large extent, the rules of the game, and how these rules are understood, applied, and also enforced.

Finally, while comparative analyses based on administrative tradition may provide insights into cross-country variation, moving between different levels of government within one country may also reveal different insights related to the dynamics of collaboration, while holding, to some extent, institutional logics constant. By exploiting the variation in administrative units (see Kuhlmann and Wollmann, 2019), it may be easier to isolate the role of project design itself or the role that the type of digitalisation may play in dictating collaboration dynamics (Wang and Feeney, 2016).

The role of administrative tradition and the desire to examine multi-levels of governance approaches is the exact approach that motivated our large-scale qualitative research project in intergovernmental collaboration. First, we selected five European countries that represent different types of administrative traditions, and also different levels of digital maturity. These countries were Belgium, Denmark, Estonia, Germany, and the United Kingdom. First, we identified key national projects related to one reform act: the EU Single Digital Gateway (SDG; The European Parliament and Council of the European Union, 2018). This approach was taken in order to add an element of control and enhance comparability between cases and also due to the necessity of collaborative efforts required to implement the act. To identify specific projects, we focused on the development of national online platforms, which are linked to the SDG act and must abide by similar guiding conditions, including following the 'once-only' principle, and 'user centricity' necessitating both vertical and horizontal collaborations. Second, based on the country-level selection, in order to hold administrative tradition somewhat constant, five municipalities were selected within these countries. Within these municipalities there were at least 50,000 inhabitants. Similar to the national level, to enhance comparability and relevance, a comparison of collaborative projects in the realm of the 'smart city' was selected due to the fact that this represents a key innovation where collaboration was necessary.

The results of this ambitious research project are extensive (Breaugh et al., 2020; Rackwitz et al., 2020), with several key themes that emerged across the case studies. First, given the complexities of digitalisation projects, especially when they cross ministerial boundaries, collaboration has become an essential, and default governance approach in tackling the complexity of the task at hand. However, in every case study, the complexity of both the content of the projects themselves, as well as managing the interpersonal relations within the collaboration were serious hurdles that needed to be overcome. Second, risk, complexity, and power did indeed become salient in each of the cases, but the large variance of when and how these dynamics emerged in the course of the collaboration suggests that no project is alike and that the public management interventions need to be specifically tailored to this context of the project itself. Third, from an institutional design perspective, the dynamics and challenges were handled primarily through a central coordinating body that managed all the projects. These central coordinating bodies held most of the steering power of the projects. However, through clear structures related to rules and processes, flatter, more inclusive, and participatory approaches emerged through the use of working groups. The clear project structures enabled clear lines of accountability (and thus managing risk), and they also acted as a means of working groups to exert power through their clearly articulated process rules and responsibilities. Finally, the role of leadership was salient in all ten case studies. Similarities in leadership styles showed the necessity for leaders who are not only sensitive to understanding these dynamics but also agile enough to handle them using different strategies. This does, without a doubt, underscore the essential role that flexible leadership plays.

CONCLUSIONS

The purpose of this chapter was to examine the collaboration dynamics of intergovernmental collaborations within the context of government digitalisation. It is clear that collaborations are an essential form of modern governance, especially when addressing the process of government digitalisation and reform. But, despite its clear necessity, it does not discount the fact that it is hard, time-intensive, and requires high levels of administrative, project, and interpersonal skills on the part of leaders. It also requires all other stakeholders to be willing and interested in engaging – without this, collaborations run the risk of being more 'talk' than 'action'. However, as the demands of a digital society increase and digital public services increase alongside this, public sector organisations have to continue to embrace collaborative approaches, breaking down silos, and accepting the new form of dispersed power and influence.

While this chapter presents the beginning stages of understanding collaboration within the context of digitalisation, there are still many questions that emerge that deserve more research attention. The first is the importance of digging deeper into the dynamics of collaboration itself. To date, except for Klijn and Koppenjan's (2016) work on understanding collaborative complexities, very few researchers have dug deeper into the different types of risk and power that emerge in collaboration projects, for example, different types of risk (related to reputation, financial and/or political) and if and how they impact collaboration behaviours differently. This would provide a more nuanced understanding of collaboration conditions and build on Ansell and Gash's (2008) collaborative conditions interdependency hypotheses. A second area of research could be to examine the notion of collaborative resistance, and why, when given the choice, organisations choose not to participate in collaboration projects. This could be related to the collaboration dynamics noted above (risk, power, and complexity), but it is also more strongly focused on practical issues related to a lack of skills or understanding of digital processes, a lack of motivation due to fear of change, or a lack of people able to engage with the process. If a lack of motivation and/or skills is the case, then training in digital skills needs to become a part of the collaboration frameworks. If it is more a lack of personnel and resources, then organisations and those heading collaboration projects must take greater consideration of the personnel costs of collaborating and account for this in their budgets. Finally, while current research focusing on collaboration has a rich understanding of how collaboration projects are set up, especially within the context of digitalisation, very little literature focuses on long-term project maintenance. Questions include aspects such as the dependency of the project managers on the digitalisation project, and if and how a handover may occur, if and how collaboration changes in the maintenance rather than set-up phase of a project (see, e.g., Brorström and Diedrich, 2022), the role of collaboration inertia in the long-term sustainability of collaboration networks (Vangen and Huxham, 2003), and how the challenges and dynamics may change as projects become more mature, goals are solidified, and the 'complexity' of the project reduces as learning and competencies increase.

Overall, the digitalisation processes that governments are currently undertaking represent a new and very large wave of reforms that will (and do) fundamentally change the way governments work and interact with citizens and other stakeholders. To this end, while digital offers technical solutions that may simplify the work of government, the re-centralisation process necessary to achieve this is not an easy task. Collaboration appears to be the best and only suitable approach to do this. While the benefits are evident, the time, energy, and highly skilled management required to collaborate effectively will continue to challenge the collaboration-first mentality for years to come.

REFERENCES

Agger, Annika, and Sørensen, E. (2018). Managing collaborative innovation in public bureaucracies. *Planning Theory*, 17(1), 53–73.

Agranoff, R., and McGuire, M. (2003). *Collaborative Public Management.* Georgetown University Press.

Ansell, C., and Gash, A. (2008). Collaborative governance in theory and practice. *Journal of Public Administration Research and Theory*, 18(4), 543–71.

Ansell, C., and Gash, A. (2012). Stewards, mediators, and catalysts: Toward a model of collaborative leadership. *Innovation Journal*, 17(1), 1–21.

Apostolou, D., Mentzas, G., Stojanovic, L., Thoenssen, B., and Pariente Lobo, T. (2011). A collaborative decision framework for managing changes in e-government services. *Government Information Quarterly*, 28(1), 101–16.

Benz, A., and Zimmerman, C. (2011). Germany: Varieties of democracy in a federal system. In J. Loughlin, F. Hendriks, and A. Lidstrom (Eds.), *The Oxford Handbook of Local and Regional Democracy in Europe* (pp. 146–72). Oxford University Press.

Bianchi, C., Nasi, G., and Rivenbark, W.C. (2021). Implementing collaborative governance: Models, experiences, and challenges. *Public Management Review*, 23(11) 1581–9.

Bingham, L.B. (2008). Legal frameworks for collaboration in governance and public management. In L.B. Bingham and R. O'Leary (Eds.), *Big Ideas in Collaborative Public Management* (pp. 247–69). M.E Sharpe.

Bommert, B. (2010). Collaborative Innovation in the public Sector. *International Public Management Review*, 11(1), 15–33.

Bovens, M. (2007). Analysing and assessing accountability: A conceptual framework. *European Law Journal*, 13(4), 447–68.

Breaugh, J., Hammerschmid, G., and Rackwitz, M. (2020). Research report on collaborative management for ICT-enabled public sector innovation. https://opus4.kobv

.de/opus4-hsog/frontdoor/index/index/searchtype/authorsearch/author/Jessica+Breaugh/docId/3994/start/8/rows/10, accessed 27 September 2023.
Brorström, S., and Diedrich, A. (2022). Boundaries of collaboration – the case of a temporary housing complex for refugees in Sweden. *Public Management Review*, 24(4), 536–57.
Brown, L., and Osborne, S.P. (2013). Risk and innovation. *Public Management Review*, 15(2), 186–208.
Chen, Y., and Lee, J. (2018). Collaborative data networks for public service: rGovernance, management, and performance. *Public Management Review*, 20(5), 672–90.
Chen, Y., Hu, L., Tseng, K., Juang, W., and Chang, C. (2019). Cross-boundary e-government systems: Determinants of performance. *Government Information Quarterly*, 36(3), 449–59.
Choi, T., and Robertson, P.J. (2014). Deliberation and decision in collaborative governance: A simulation of approaches to mitigate power imbalance. *Journal of Public Administration Research and Theory*, 24(2), 495–518. https://doi.org/10.1093/jopart/mut003.
Clarke, A. (2019). Digital government units: What are they, and what do they mean for digital era public management renewal? *International Public Management Journal*, 23(3), 358–79.
Clausen, T., Demircioglu, M., and Alsos, G. (2020). Intensity of innovation in public sector organizations: The role of push and pull factors. *Public Administration*, 98, 159–76.
Cristofoli, D., Meneguzzo, M., and Riccucci, N. (2017). Collaborative administration: The management of successful networks. *Public Management Review*, 19(3), 275–83.
Cohen, G., and Cohen, N. (2021). Understanding street-level bureaucrats' informal collaboration: Evidence from police officers across the jurisdictional divide. *Public Management Review*, 1–19. https://doi.org/10.1080/14719037.2021.1963824.
Crosby, B., and Bryson, J. (2010). Integrative leadership and the creation and maintenance of cross-sector collaborations. *Leadership Quarterly*, 21(2), 211–30.
Dawes, S., and Pardo, T. (2002). Building collaborative digital government systems. In W. McIver and A. Elmagarmid (Eds.), *Advances in Digital Government* (Vol. 26, pp. 259–73). Springer.
Dawes, S., Cresswell, A., and Pardo, T. (2009). From 'need to know' to 'need to share': Tangled problems, information boundaries, and the building of public sector knowledge networks. *Public Administration Review*, 69(3), 392–402.
de Vries, H., Bekkers, V., and Tummers, L. (2016). Innovation in the public sector: A systematic review and future research agenda. *Public Administration*, 94(1), 146–66.
Dearborn, D., and Simon, H. (1958). Selective perception: A note on the departmental identifications of executives. *Sociometry*, 21(2), 140–4.
Dickinson, H., and Sullivan, H. (2014). Towards a general theory of collaborative performance: The importance of efficacy and agency. *Public Administration*, 92(1), 161–77.
Doberstein, C. (2016). Designing collaborative governance decision-making in search of a 'collaborative advantage.' *Public Management Review*, 18(6), 819–41.
Elnaghi, M., Alshawi, S.N., Kamal, M.M., Weerakkody, V., and Irani, Z. (2019). Exploring the role of government authority in managing transformation in service re-engineering – experiences from Dubai police. *Government Information Quarterly*, 36(2), 196–207.

Emerson, K., Nabatchi, T., and Balogh, S. (2012). An integrative framework for collaborative governance. *Journal of Public Administration Research and Theory*, 22(1), 1–29.

Eriksson, E., Andersson, T., Hellström, A., Gadolin, C., and Lifvergren, S. (2020). Collaborative public management: Coordinated value propositions among public service organizations. *Public Management Review*, 22(6), 791–812.

Fedorowicz, J., Sawyer, S., Williams, C.B., Markus, M.L., Dias, M., Tyworth, M., Gantman, S., Jacobson, D., Tomasino, A.P., and Schrier, R. (2014). Design observations for interagency collaboration. *Government Information Quarterly*, 31(2), 302–16.

Ferro, E., and Sorrentino, M. (2010). Can intermunicipal collaboration help the diffusion of e-government in peripheral areas? Evidence from Italy. *Government Information Quarterly*, 27(1), 17–25.

Fountain, J.E. (2001). *Building the Virtual State: Information Technology and Institutional Change*. Brookings Institution Press.

Gil-García, J.R., Chengalur-Smith, I., and Duchessi, P. (2007). Collaborative e-government: Impediments and benefits of information-sharing projects in the public sector. *European Journal of Information Systems*, 16(2), 121–33. https://doi.org/10.1057/palgrave.ejis.3000673.

Gil-García, J., Dawes, S., and Pardo, T. (2018). Digital government and public management research: Finding the crossroads. *Public Management Review*, 20(5), 633–46.

Gil-García, J., Guler, A., Pardo, T., and Burke, G. (2019). Characterizing the importance of clarity of roles and responsibilities in government inter-organizational collaboration and information-sharing initiatives. *Government Information Quarterly*, 36(4), 101393.

Guy Peters, B., and Pierre, J. (2019). Populism and public administration: Confronting the administrative state. *Administration and Society*, 51(10), 1521–45.

Hartley, J. (2005). Innovation in governance and public services: Past and present. *Public Money and Management*, 25(1), 27–34. https://doi.org/10.1111/J.1467-9302.2005.00447.X.

Hartley, J., Sørensen, E., and Torfing, J. (2013). Collaborative innovation: A viable alternative to market competition and organizational entrepreneurship. *Public Administration Review*, 73(6), 821–30.

Head, B. (2008). Assessing network-based collaborations: Effectiveness for whom? *Public Management Review*, 10(6), 733–49.

Hood, C. (1991). A public management for all seasons? *Public Administration*, 69, 3–19.

Huxham, C., and Vangen, S. (2005). *Managing to Collaborate: The Theory and Practice of Collaborative Advantage*. Routledge.

Huxley, K., Andrews, R., Downe, J., and Guarneros-Meza, V. (2016). Administrative traditions and citizen participation in public policy: A comparative study of France, Germany, the UK and Norway. *Policy and Politics*, 44(3), 383–402.

Jaeger, P. (2002). Constitutional principles and e-government: An opinion about possible effects of federalism and the separation of powers on e-government policies. *Government Information Quarterly*, 19(4), 357–68.

Jensen, U., Andersen, L., Bro, L., Bøllingtoft, A., Eriksen, T., Holten, A., Jacobsen, C., Ladenburg, J., Nielsen, P., Salomonsen, H., Westergård-Nielsen, N., and Würtz, A. (2019). Conceptualizing and measuring transformational and transactional leadership. *Administration and Society*, 51(1), 3–33.

Jones, S., and Hooper, T. (2017). New Zealand's ICT strategy: The respective roles of senior and middle management in promoting collaboration and innovation. *Asia and the Pacific Policy Studies*, 4(3), 484–95.

Kickert, W., Klijn, E., and Koppenjan, J. (1997). *Managing Complex Networks*. Sage.

Klijn, E., and Edelenbos, J. (2007). Theories of democratic network governance. In E. Sørensen and J. Torfing (Eds.), *Theories of Democratic Network Governance* (pp. 199–214). Palgrave Macmillan.

Klijn, E., and Koppenjan, J. (2000). Public management and policy networks. *Public Management: An International Journal of Research and Theory*, 2(2), 135–58.

Klijn, E., and Koppenjan, J. (2014). Complexity in governance network theory. *Complexity, Governance and* Networks, 1(1), 61–70.

Klijn, E., and Koppenjan, J. (2016). *Governance Networks in the Public Sector*. Routledge.

Krogh, A.H. (2022). Facilitating collaboration in publicly mandated governance networks. *Public Management Review*, 24(4), 631–53.

Krogh, A., and Torfing, J. (2015). Leading collaborative innovation: Developing innovative solutions to wicked gang problems. In A. Agger, B. Damgaard, A. Hagedorn Krog, and E. Sørensen (Eds.), *Collaborative Governance and Public Innovation in Northern Europe* (pp. 91–110). Bentham Science Publishers.

Kuhlmann, S., and Wollmann. (2019). *Introduction to Comparative Public Administration: Administrative Systems and Reforms in Europe*. Edward Elgar Publishing.

Lasker, R., Weiss, E., Baker, Q., Collier, A., Israel, B., Plough, A., and Bruner, C. (2003). Broadening participation in community problem solving: A multidisciplinary model to support collaborative practice and research. *Journal of Urban Health*, 80(1), 14–47.

Lewis, J., Ricard, L., and Klijn, E. (2018). How innovation drivers, networking and leadership shape public sector innovation capacity. *International Review of Administrative Sciences*, 84(2), 288–307.

Liu, X., and Zheng, L. (2018). Cross-departmental collaboration in one-stop service center for smart governance in China: Factors, strategies and effectiveness. *Government Information Quarterly*, 35(4), S54–S60.

Luna-Reyes, L., Gil-García, J., and Cruz, C. (2007). Collaborative digital government in Mexico: Some lessons from federal web-based interorganizational information integration initiatives. *Government Information Quarterly*, 24(4), 808–26.

Margetts, H., and Dunleavy, P. (2013). The second wave of digital-era governance: A quasi-paradigm for government on the Web. *Philosophical Transactions of the Royal Society A Mathematical, Physical and Engineering Sciences*, 371, 1–17.

McGuire, M. (2006). Collaborative public management: Assessing what we know and how we know it. *Public Administration Review*, 66(Suppl. 1), 33–43.

McGuire, M., and Agranoff, R. (2011). The limitations of public management networks. *Public Administration*, 89(2), 265–84.

Mergel, I. (2016). Agile innovation management in government: A research agenda. *Government Information Quarterly*, 33(3), 516–23.

Mergel, I. (2018). Open innovation in the public sector: Drivers and barriers for the adoption of Challenge.gov. *Public Management Review*, 20(5), 726–45.

Mergel, I., Gong, Y., and Bertot, J. (2018). Agile government: Systematic literature review and future research. *Government Information Quarterly*, 35(2), 291–8.

Mergel, I., Edelmann, N., and Haug, N. (2019). Defining digital transformation: Results from expert interviews. *Government Information Quarterly*, 36(4), 101385.

Meuleman, L. (2008). *Public Management and the Metagovernance of Hierarchies, Networks and Markets*. Physica-Verlag HD.

Meijer, A. (2015). E-governance innovation: Barriers and strategies. *Government Information Quarterly*, 32(2), 198–206.

Meyer, R., and Hammerschmid, G. (2010). The degree of decentralization and individual decision making in central government human resource management: A European comparative perspective. *Public Administration*, 88(2), 455–78.

Moore, M. (1995). *Creating Public Value*. Harvard University Press.

Morse, R. (2010). Integrative public leadership: Catalyzing collaboration to create public value. *The Leadership Quarterly*, 21(2), 231–45.

Neumann, O., Matt, C., Hitz-Gamper, B.S., Schmidthuber, L., and Stürmer, M. (2019). Joining forces for public value creation? Exploring collaborative innovation in smart city initiatives. *Government Information Quarterly*, 36(4), 101411.

O'Leary, R., Gerard, C., and Choi, Y. (2012). The skill set of the successful collaborator. *Public Administration Review*, 72(S1), 70–83.

Picazo-Vela, S., Gutiérrez-Martínez, I., Duhamel, F., Luna, D., and Luna-Reyes, L. (2018). Value of inter-organizational collaboration in digital government projects. *Public Management Review*, 20(5), 691–708.

Provan, K., and Kenis, P. (2008). Modes of network governance: Structure, management, and effectiveness. *Journal of Public Administration Research and Theory*, 18(2), 229–52.

Purdy, J.M. (2012). A framework for assessing power in collaborative governance processes. *Public Administration Review*, 72(3), 409–17.

Rackwitz, M., Hammerschmid, G., Breaugh, J., and Palaric, E. (2020). Government collaboration and digitalisation: Comparative case studies on collaboration management for government digitalisation and public sector innovation. https://tropico-project.eu/download/d6-3-government-collaboration-and-digitalisation-comparative-case-studies-on-collaborative-management-for-government-digitalisation-and-public-sector-innovation/, accessed 27 September 2023.

Ran, B., and Qi, H. (2018). Contingencies of power sharing in collaborative governance. *American Review of Public Administration*, 48(8), 836–51.

Ran, B., and Qi, H. (2019). The entangled twins: Power and trust in collaborative governance. *Administration and Society*, 51(4), 607–36.

Ricard, L., Klijn, E., Lewis, J., and Ysa, T. (2017). Assessing public leadership styles for innovation: A comparison of Copenhagen, Rotterdam and Barcelona. *Public Management Review*, 19(2), 134–56. https://doi.org/10.1080/14719037.2016.1148192.

Rittel, H.W. and Webber, M.M. (1973). Dilemmas in a general theory of planning. *Policy Sciences*, 4(2), 155–69.

Ruggieri, S., and Abbate, C.S. (2013). Leadership style, self-sacrifice, and team identification. *Social Behavior and Personality: An International Journal*, 41(7), 1171–8.

Ruggieri, S., and Abbate, C.S. (2013). Leadership style, self-sacrifice, and team identification. *Social Behavior and Personality: An International Journal*, 41(7), 1171–8.

Scholl, H., Kubicek, H., Cimander, R., and Klischewski, R. (2012). Process integration, information sharing, and system interoperation in government: A comparative case analysis. *Government Information Quarterly*, 29(3), 313–23.

Scott, R.J., and Merton, E.R.K. (2021). When the going gets tough, the goal-committed get going: Overcoming the transaction costs of inter-agency collaborative governance. *Public Management Review*, 23(11), 1–24.

Sedgwick, D. (2016). Building collaboration: Examining the relationship between collaborative processes and activities. *Journal of Public Administration Research and Theory*, 27(2), 236–52.

Span, K., Luijkx, K., Schalk, R., and Schols, J. (2012). What governance roles do municipalities use in Dutch local social support networks? *Public Management Review*, 14(8), 1175–94.

Termeer, C., and Dewulf, A. (2019). A small wins framework to overcome the evaluation paradox of governing wicked problems. *Policy and Society*, 38(2), 298–314.

The European Parliament, and Council of the European Union. (2018). *Regulation (EU) 2018/1724 of the European Parliament and of the Council of 2 October 2018 establishing a single digital gateway to provide access to information, to procedures and to assistance and problem-solving services and amending Regulation (EU) No. 1024/2012.* Official Journal of the European Union. https://eur-lex.europa.eu/legal-content/EN/TXT/?uri=uriserv:OJ.L_.2018.295.01.0001.01.ENG.

Timeus, K. (2019). Passing the buck? How risk behaviours shape collaborative innovation. In T. Bach and K. Wegrich (Eds.), *The Blind Spots of Public Bureaucracies and the Politics of Non-Coordination*. Palgrave Macmillan.

Torfing, J. (2019). Collaborative innovation in the public sector: The argument. *Public Management Review*, 21(1), 1–11.

Torfing, J., Sørensen, E., and Bentzen, T.O. (2019). Institutional design for collective and holistic political leadership. *International Journal of Public Leadership*, 15(1), 58–76.

Van Wart, M. (2003). Public-sector leadership theory: An assessment. *Public Administration Review*, 63(2), 214–28.

Vangen, S., and Huxham, C. (2003). Nurturing collaborative relations: Building trust in interorganizational collaboration. *The Journal of Applied Behavioral Science*, 39(1), 5–31.

Wang, S., and Feeney, M. (2016). Determinants of information and communication technology adoption in municipalities. *American Review of Public Administration*, 46(3), 292–313.

Wegrich, K. (2019). The blind spots of collaborative innovation. *Public Management Review*, 21(1), 12–20.

4. Conditions for successful public-private collaboration for public service innovation

Chesney Callens and Koen Verhoest

INTRODUCTION

Digital transformation and public service innovation are being propelled by partnerships that unite public sector actors (e.g., governments, agencies, public hospitals, etc.) and private sector actors (e.g., third sector organizations, firms, grassroots organizations, etc.). The basic premise to use these types of configurations is that today's societal problems have become too large, complicated, and interconnected to expect solutions from individual organizations (Sørensen and Torfing, 2020). Public sector organizations are exploring different types of collaborative arrangements such as public-private partnerships (PPPs) (Brogaard, 2017; Callens et al., 2021), triple-helix partnerships, which include universities (Leydesdorff and Meyer, 2003), and public-private innovation partnerships (PPIs) (Brogaard, 2021; Di Meglio, 2013) to pursue public service innovation. The search for public service innovation is, however, often entangled with digital transformation initiatives, as disruptive technological innovations are a crucial part of digital transformation (Nadkarni and Prügl, 2021). In complex policy sectors, digital transformation through technological innovation can often only be achieved through collaborative partnerships, as no individual actor possesses all the required knowledge, resources, and capabilities to innovate. Furthermore, these partnerships are often *public-private* collaborations, as government organizations have important political incentives, regulatory powers, and public resources that can be directed towards innovation, while private sector organizations possess the specialized knowledge and capabilities to develop these innovations. Hence, partnership synergies emerge between public and private actors (Lasker et al., 2001), which can lead to disruptive and transformative public service innovations.

The purpose of this chapter is to develop a theoretical framework that unpacks structure and agency-related conditions of public-private collaborations that are directed towards the creation and implementation of digital service innovations. The theoretical framework not only allows us to conceptually and analytically study the design and agency of such partnerships but also to gauge the effect of these elements on the success and innovativeness of these partnerships. The framework connects four different types of conditions, which influence the process of collaborative innovation. The first set of conditions focuses on the features of the collaboration at the level of the partnership. We explain how structure-related conditions, such as the partnership composition and governance structure, and agency-related conditions, such as the management of the partnership, can affect the innovation process. The second set of conditions considers features at the level of the individuals and organizations involved in the partnership. Conditions such as interpersonal trust, knowledge and skills, and external support are considered. The third set of conditions focuses on how technological structures such as the use of information and communication technology (ICT) affect the collaborative innovation process. ICTs are a central characteristic of digital transformation, but ICT use can also be an enabler for enhanced collaborative innovation, because of their impact on collaborative dynamics between partners, and on the service design. Last but not least, the fourth set of conditions is directed towards the involvement of service users in the innovation process. Users are crucial agents in innovation processes, as they can legitimate, support, and even drive the innovation process. We consider various features and conditions of user involvement.

DIGITAL TRANSFORMATION THROUGH PUBLIC SERVICE INNOVATION

Digital transformation is related to the changes organizations, sectors, and societies undergo as a result of the introduction and implementation of digital technologies (Vial, 2019). An important way to obtain digital transformation is through the adoption of public service innovation. Indeed, through the use of highly advanced digital service innovations, governments are able to transform their processes, routines, work tools, and service delivery (Mergel et al., 2019). However, digital transformation goes further than its impact on government organizations, as digital innovations have often a broad impact on industries, governance structures, and policy ecosystems (Eom and Lee, 2022). Public service innovations, such as smart city technologies, eHealth technologies, and COVID-19 response technologies affect many societal stakeholders (i.e., private companies, non-profits, public sector organizations, citizens and users, etc.), and have dramatically changed important aspects of our societies. For instance, digital transformation in healthcare affects the quality of healthcare (Agarwal et al., 2010), but also the health standard of communities and societies as new treatments become possible or crucial health information becomes accessible to healthcare providers. Hence, digital service innovation propels the digital transformation of significant parts of our society.

Although these digital public service innovations have a huge impact on our society, even now, innovation in the public sector is being exposed to much scepticism (Hartley, 2005), and, until recently, the word 'public sector innovation' was regarded as an oxymoron (Bommert, 2010; Torfing et al., 2020). While the concept of innovation as a research subject emerged from 'Schumpeterian' economics in the private sector, in recent decades, innovation has been intensively researched in the public sector as well. Indeed, evidence from the public sector shows how governments are often responsible for important technological breakthroughs, with classic examples such as the invention of the World Wide Web and biotechnological innovations (Windrum, 2008), but also that governments are at least equally proficient at organizational innovation as many private sector organizations (Djellal et al., 2013; Earl, 2004; Windrum, 2008). Public service innovation is regarded as a means through which complex societal issues can be solved, the rising demands of citizens can be achieved, and government resources can be spent more efficiently (de Vries et al., 2018).

Digital service innovations are digital services that are 'perceived as new by an individual or other unit of adoption' (Rogers, 2003, p. 12). The process by which these services arise is crucial to understand how organizational innovation can be stimulated. The innovation process is defined by Damanpour and Schneider (2008, p. 496) as 'the development (generation) and/or use (adoption) of new ideas or behaviours'. Two important components of the innovation process can be distinguished from this definition. First, during the idea generation phase of the innovation process, ideas are proposed, circulated, discussed, integrated, transformed, and selected by the innovators. Second, during the implementation phase of the innovation process, the selected ideas are translated into implementable digital solutions, which can be practically adopted by users. Idea testing can work as a gateway between idea generation and idea implementation, as selected ideas that are tested might either be eligible for idea implementation or, if they are not, they can be circled back to the idea generation phase of the innovation process (Meijer, 2014). This circular motion of the innovation process is illustrated in Figure 4.1. Note that the proposed characteristics of the innovation process are widely shared amongst innovation scholars in the public sector. Innovation scholars such as Walker (2007), Sørensen and Torfing (2011), de Vries et al. (2015), and Cinar et al. (2019) recognize similar phases of the innovation process. However, scholars also emphasize that innovation processes are intrinsically chaotic, and the phases might therefore overlap with each other (Meijer, 2014).

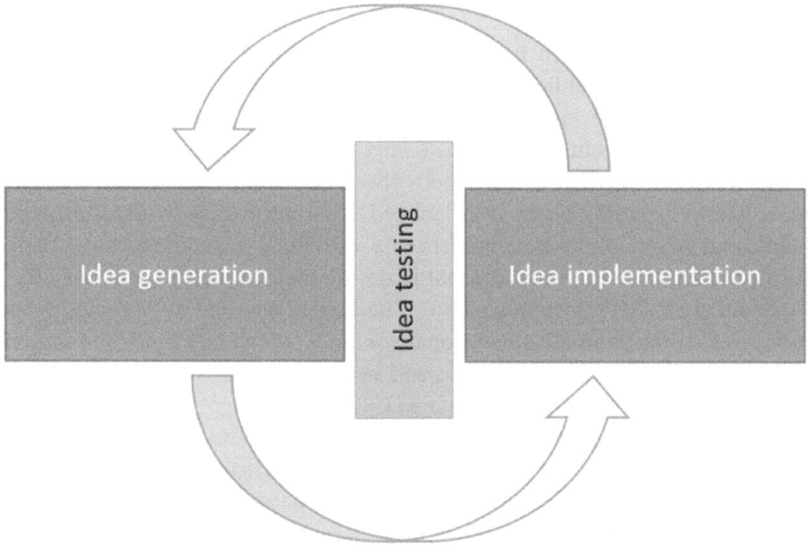

Figure 4.1 *The innovation process*

Although the literature on public service innovation has expanded dramatically in the last decades, there are still some challenges and drawbacks to public service innovation that need a proper discussion. First, critiques of innovation in the public sector argue that the government is often not suited for innovation because of its rule-bound, command-and-control, bureaucratic nature (Hartley et al., 2013). Second, in contrast to the private sector, where the concept of innovation was introduced by Schumpeter (1942), the public sector is less sensitive to or affected by competition dynamics. Innovation can be regarded as an optimization strategy in the public sector, while it is a survival strategy in the private sector (cf. 'creative destruction', Schumpeter, 1942). Even then, as governments are not directly punished with decreased revenues when their performance should go down, they are also not incentivized to use this optimization strategy (Gullmark, 2021). Third, governments are more risk-aversive than private sector organizations, as they use public resources that are externally controlled by politicians, the media, and the public (Gullmark, 2021). Fourth, innovation has a large disruptive potential, which is not always desired in the public sector (Wynen et al., 2020). As most government organizations have a monopolistic position, pursuing innovations that are simultaneously highly disruptive and very unpredictable, and risky, might endanger the continuity of public policy and service delivery.

COLLABORATIVE INNOVATION

In order to meet some of these critiques regarding innovation in the public sector, the field of public sector collaborative innovation has been growing exponentially in the last decades. On the one hand, service innovation is viewed as an important way to solve complex, wicked problems, which cannot be solved through traditional methods (Sørensen and Torfing, 2011). On the other hand, the public sector is influenced by an intricate network of different public and private sector stakeholders, and individual stakeholders have, therefore, problems solving these societal issues on their own (Bryson et al., 2015; Emerson et al., 2011; Osborne, 2006). The current complexity of societies and industries demands multi-dimensional solutions which can transcend organizational boundaries, policy sectors, and even spheres of societies (Crosby et al., 2017; Diamond and Vangen, 2017). For this reason, public and private sector stakeholders work together in networks and partnerships, from which new policies and services can emerge (Ansell and Gash, 2007; Bryson et al., 2006). These public-private collaborations are potentially valuable breeding grounds for innovation, as partnership synergies might arise from the close interactions between the involved actors (Lasker et al., 2001). Knowledge, resources, and perspectives from a diverse set of stakeholders are integrated in those partnerships, which might result in the adoption or creation of new frames and ideas, from which innovations might arise (Torfing, 2019). Recent research by Wilson and Mergel (2022) confirms the advantages of 'networking strategies' for overcoming structural and cultural barriers to digital transformation. The majority of the interviewed digital government champions in their study indicated that networks and collaboration enable the opportunity to learn from peers and exchange knowledge across contexts. Furthermore, collaborations are better able to share the risks and costs of creating and implementing innovation, which would otherwise all end up in the same organization (Appleyard and Chesbrough, 2017).

Collaborative innovation has several advantages over other forms of innovation (e.g., in-house innovation). We argue that these advantages can be summarized into two principal premises of collaborative innovation (see also Callens et al., 2020). First, collaboration has a *reinvigorating* effect on the innovation process. Collaborative innovation allows partners from different backgrounds to access new knowledge, connect and build on each other's ideas, and adopt new perspectives, which stimulates divergent thinking and generates creative momentum (Paulus et al., 2018). Collaborative innovation opens up the innovation process to a broader ideation context from which new ideas can arise more easily. By opening up the innovation process, collaborative innovation prevents the innovation process from being trapped in the convergent thinking of like-minded people, which could lead to groupthink and tunnel vision, and which inhibits innovation (Torfing et al., 2020). Second, collaboration also has a *protective* effect on the innovation process. Indeed, through collaborative innovation, difficult problems such as wicked issues are shared amongst multiple stakeholders (Crosby et al., 2017), as are the risks and costs that are tied to inventing something new (Corsaro et al., 2012). Protecting the innovation process with the resources and commitment of multiple stakeholders is no unnecessary precaution, as innovations have a high chance of failure (Van der Panne et al., 2003). Furthermore, innovations often require significant investments, but do not always translate properly to a real-life context, nor are they always sufficiently adopted by the public or sustainable in the long run (Brown and Osborne, 2013, p. 187). In collaborative innovation, multiple stakeholders are responsible for creating and implementing the innovation, which increases their capacity to avert innovation failures. Furthermore, the collaboration might also create an isolated, socio-technical niche in which experimentation and trial-and-error behaviour are tolerated, without the risk of falling prey to highly competitive (market) dynamics (Hermans et al., 2013).

Sørensen and Torfing (2011) and Ansell and Torfing (2014) propose four interrelated processes of collaborative innovation. First, public-private collaborations integrate knowledge, perspectives, and resources from a variety of stakeholders, which increases the likelihood that *synergies* arise. Such partnership synergies are important for the quality of the interaction in collaborations (Lasker et al., 2001) but are also essential for the creation of something new, as new ideas can arise out of the combination of different knowledge pools and perspectives, and collective capacity can be activated by sharing and connecting resources and skills from multiple actors (Waldorff et al., 2014). Second, by interacting with each other, individuals exchange information and knowledge, which allows them to create new associations between distinct concepts and learn from each other (Ansell and Torfing, 2014). *Learning* is an important second process of collaborative innovation because it uses the available variance in the partnership and transforms it into novel and creative ideas, for instance, by building on other's knowledge and ideas (Hartley and Rashman, 2018). Third, *consensus building* allows the partners to arrive at a shared idea, towards which the efforts of the partners can be directed. Through consensus building, the partners can search for agreement and similarities between perspectives (Innes and Booher, 1999), which should eventually lead to joint ownership over the idea (Lindsay et al., 2020). Fourth, building *commitment* to implement the idea is a final process of collaborative innovation (Trivellato et al., 2021). Commitment refers to the willingness of the involved stakeholders to mobilize resources towards the implementation of the innovation.

However, as with public service innovation itself, collaborative innovation also has some drawbacks that need to be recognized. These drawbacks are particularly related to the inherent tension between creative ideation and collaborative stability (Torfing, 2019), and the inefficiencies surrounding collaboration. Collaboration is often a lengthy and underperforming process, because of the consistent need for aligning and realigning different perspectives, visions, interests, etc. (Huxham, 2003). Klijn and Koppenjan (2015) discuss three types of network complexities that are common in partnerships (i.e., substantive complexities, strategic complexities, and institutional complexities), which need to be held in check if the partnership wants to optimize its performance. Several network management strategies are proposed by the authors to manage these complexities (i.e., exploring, connecting, arranging, processing rules). However, managing these complexities means that coordinators need to invest a lot of time into the network interactions, which increases the transaction costs, and potentially lowers its performance compared to other organizational arrangements (Jobin, 2008). Furthermore, as innovation thrives on the variance that is introduced in these partnerships (Sørensen and Torfing, 2017), collaborative innovation should increase these complexities and related transaction costs even further. In other words, the pursuit of innovation through collaboration might magnify the drawbacks of public-private collaborations. Collaborating for technological innovation might be even more challenging because of the technical complexity of the content of the innovation process, and the required variety of resources and knowledge in such partnerships (Picazo-Vela et al., 2018).

From these arguments, it becomes clear that pursuing digital transformation through collaborative innovation requires a prudent approach. Too much focus on conditions that stimulate variance and creative ideation might result in the premature termination of the partnership because of the increasing complexities and transaction costs. However, too much focus on reducing the transaction costs and complexities might extinguish any creative upsurge and increase the risk of groupthink. In the following section, we explore which conditions contribute to this delicate balance. Some of these conditions will be aimed at controlling and managing the collaborative innovation process (e.g., partnership structure, management, etc.), while other conditions will be focused on increasing the variance and creative potential of the partnership (e.g., knowledge and skills of partners, user involvement, etc.).

PUBLIC-PRIVATE INNOVATION PARTNERSHIPS

As mentioned in the introduction, multiple types of public-private collaborations can exist. We will focus in this chapter on public-private innovation partnerships (PPIs), which are partnerships between public actors and private actors that are aimed at producing innovative services, for which they often involve service users (Brogaard, 2021). These types of partnerships are relatively short-term partnerships, in which not only commercial firms are involved, but also non-profit or third-sector organizations (Di Meglio, 2013). In contrast to, for instance, public-private partnerships (PPPs), PPIs are less formalized, and the public partner (e.g., government) often adopts a 'leading role as initiators, organisers and propagators of new ideas' (Di Meglio, 2013, p. 80).

CONDITIONS OF COLLABORATIVE INNOVATION

Figure 4.2 shows the conceptual model that we use in this chapter, which was part of the Horizon 2020 TROPICO research project.[1] Building on a broad variety of research, including literature on collaborative innovation research, public service innovation, collaborative governance and coproduction, the conceptual model integrates several types of conditions. We propose that these conditions stimulate the processes of collaborative innovation, which were discussed in the previous section. These processes of collaborative innovation enable public-private collaborations to generate innovative public services. While we recognize that *ex ante* (e.g., 'starting conditions', Ansell and Gash, 2007) and *ex post* conditions (e.g., diffusion-related conditions, Rogers, 2003) might also influence collaborative innovation, this chapter focuses particularly on the conditions during the process of collaborative innovation. We consider four clusters of conditions, that is, conditions on the level of the partnership, conditions on the level of the involved individuals and organizations, conditions related to ICT, and conditions related to user involvement. This section discusses these clusters of conditions.

As indicated in Figure 4.2, the first cluster includes features of the partnerships. We consider particularly structural partnership features and management-related features. For the structural partnership features, we focus on the composition of the partnership and the governance structure that is used in the partnership. With regard to the management-related conditions, we consider two types of management, that is, contract management and network management. The second cluster is composed of conditions on the level of the individual partners and involved organizations. We consider conditions such as the interpersonal trust between the involved individuals and organizations, the knowledge and skills of these individuals, and support of the external environment (e.g., policy sector, politicians, media). The third cluster introduces ICT-related conditions. We focus on multiple ways in which ICT can contribute to the collaborative innovation process, both internally (i.e., by stimulating collaborative interactions) and externally (i.e., by creating an enabling environment). The fourth cluster considers how users are involved in processes of collaborative innovation. Users are amongst the most important stakeholders in innovation projects because of their knowledge of and experiences in the

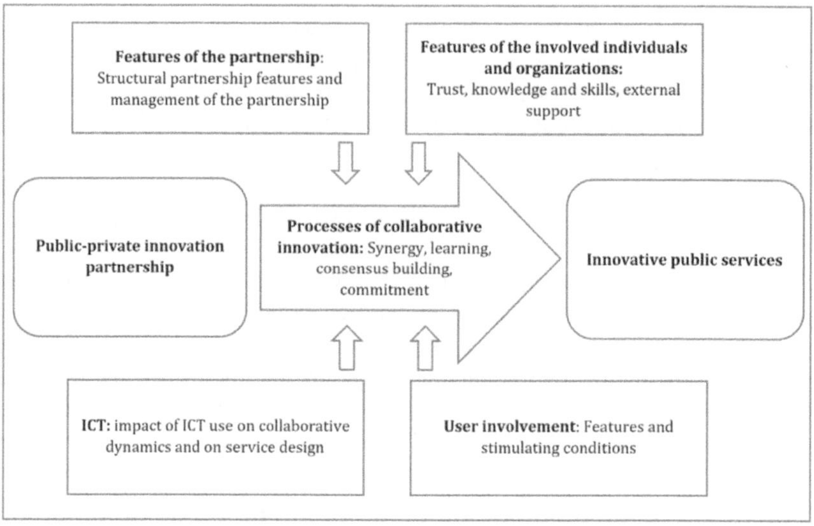

Figure 4.2 Conceptual framework

service context. We consider how different features and conditions of user involvement may affect the process of collaborative innovation.

FEATURES OF THE PARTNERSHIP

Partnership Structure

As we discussed in the previous sections, the presence of some initial variance in the partnership is necessary to kick off the innovation process. However, too much variance can lead to increased network complexities and transaction costs. This duality makes the selection of the partners in the collaboration extremely important. Partnerships with stakeholders that are very similar to each other might not produce enough synergies and learning opportunities to initiate the innovation process, while partnerships with stakeholders that are very different from each other might have difficulties working together and building trust, which can lead to the collapse of the collaboration (Torfing, 2019).

Furthermore, the composition of the partnership will also be dependent on the objectives of the partnership. Sørensen and Torfing (2017) argue that the selection of partners will differ if the goal of the partnership is to enhance the efficiency and effectiveness of public governance, enhance the democratic legitimacy of public sector organizations, or create public sector innovation.

If the partnership is established to enhance the efficiency and effectiveness of public governance (e.g., increasing the coordination between disparate public and private organizations), the authors argue that the partnership should include actors with resources and capabilities that are relevant to ensure optimal use of existing resources. A partnership that wants to enhance the democratic legitimacy of certain organizations or institutions, should, according to the authors, include a broad selection of representatives of the affected organizations or institutions (e.g., interest groups). Following Sørensen and Torfing (2017), partnerships that want to pursue public sector innovation should include a wide variety of stakeholders that possess innovation capabilities (e.g., expert knowledge, field experience, creative thinking, testing and implementation capabilities, etc.).

Note, however, that these three types of objectives of partnerships will often be intertwined in processes of collaborative innovation. Partnerships that want to pursue innovation will most likely also have to engage a wide selection of representatives, as public innovations will have consequences for public policy and service delivery, and are often part of or connected to government programmes. Furthermore, innovations are often aimed at introducing systems that allow more efficient and effective public governance, which means that stakeholders who can optimize the use of existing resources (e.g., by connecting disparate resources) will probably also be involved in innovation-oriented partnerships.

Governance Structure

The governance structure of a partnership is crucial for the interaction dynamics that unfold during the collaboration process. Governance structures build a framework around collaborative interactions, which determines which actors are included in the partnership, which responsibilities the involved actors have, how these actors are positioned towards each other, and how decisions are made (Lopes and Farias, 2022). For instance, in innovation partnerships, user involvement might be of particular importance because they are part of the context in which the innovation will eventually be adopted (see later). Through a governance structure, the partnership is able to structurally involve the users by, for instance, establishing a project team of which the users are part. This allows a more inclusive, transparent, and empowered engagement of the users, which might stimulate intensive user involvement throughout the whole innovation process. Furthermore, the governance structure of a partnership connects the resource support of the represented organizations (e.g., governments or private sector financers) with the bottom-up innovation initiatives (Lam and Li, 2018). Through various structures (e.g., steering committees, project teams, work groups, etc.), the partnership introduces a semi-hierarchical system that allows proper coordination, management, and accountability of the partnership's activities, but also protects the creative and experimental approach of the innovation process.

Provan and Kenis (2007) distinguish three types of governance structures. The first governance structure is the 'shared participant-governed' structure, which involves the equal involvement and contribution of all the involved actors in the partnership. All involved actors are responsible for the governance of the partnership and have equal decision power. The authors suggest using this type of governance structure when there is a limited number of partners, a lot of trust and goal consensus between the partners, and a low need for specialized network management skills. The second governance structure is the 'lead organization-governed' structure, which entails that one or a few actors are responsible for the governance of the partnership and have high levels of decision power. The lead actor functions as a central broker in the partnership and has the power to enforce decisions. According to the authors, this governance structure should be used in partnerships with a moderate number of participants, low levels of trust and goal consensus between the partners, and a moderate need for specialized network management skills. The third governance structure the authors propose is the 'network-administrative organisation'. This type of partnership structure establishes a separate administrative entity that is responsible for the network governance and acts as a broker. The authors advise using this type of governance structure in partnerships that have a large number of actors, a moderate level of trust and goal consensus between the partners, and need a lot of specialized network management skills (e.g., because of the presence of a lot of network complexities).

Note that there is no consensus in the literature on which type of governance structure is more suited to produce collaborative innovation. For instance, while high levels of goal consensus (e.g., in shared participant-governed partnerships) can stimulate collaborative interactions between the partners, it might also reduce divergent thinking and create tunnel vision. Similarly, the centralization of decision power (e.g., in lead organization-governed partnerships) might reduce open interaction and free circulation of ideas and perspectives, but it might also reduce the interaction costs that arise from network complexities. Some tentative results from recent research indicate that more centralized partnerships such as lead organization-governed partnerships are actually better suited for collaborative innovation (Lam and Li, 2018). However, future research should take a closer look at how different governance structures affect the collaborative innovation process.

Partnership Management

Contract management

Innovation in partnerships can be stimulated by managing the contractual foundations that underlie the collaboration. Indeed, many innovation partnerships have one or more (formal or informal) agreements between the involved partners that comprise what the partnership wants to achieve. Research into public procurement for innovation gives us insights into how contract management can stimulate innovation. Contract management is a demand-side instrument, which means that a public procurer demands a certain innovation from one or more contractors through conditions that are stipulated in a contract (Callens et al., 2021), which increases the control over the accomplishment of certain demands (Edquist et al., 2015). As innovation and collaborations are inherently risky and unpredictable, public procurers can formulate conditions in contracts, which protect them from undesired project deviations. Through contract management, the public procurer is able to share the risks of failure with the involved partners and reduce the transaction costs that are connected to the unpredictability of collaborative innovation processes. Indeed, demands that are stipulated in a written agreement can be legally enforced by the procurer, which gives the procurer a lot of control over the end result of the collaboration. This helps to increase the commitment of the partners to develop and implement the requested end product. Furthermore, the contract makes the objectives and conditions of the collaboration clear for every involved actor, which increases the goal consensus between the partners and eases the collaboration.

Contract management can be exercised through several instruments, which are listed by Uyarra et al. (2014). The contract instruments that are most often used are the inclusion of innovation-oriented output specification in the contract, innovation incentives in the tender award criteria during the procurement process, design freedom or contract flexibility that allow changes in the contract that are caused by the unpredictable nature of innovation process, and contract sanctions in case the partners deviate from what was agreed (Edler and Georghiou, 2007; Georghiou et al., 2014; Leiringer, 2006; Tadelis and Bajari, 2006). All these instruments direct the partners towards the expected outcome of the collaborative innovation process. Some of the instruments have a direct influence on the pursuit of innovation, for instance, by specifying what needs to be accomplished through output specifications or by selecting more innovative proposals through the use of innovation-oriented tender award criteria. Other instruments work indirectly on the contract partners, for instance, by allowing some flexibility in the contract in order to redirect the project when new information is obtained (e.g., new direction due to the results of a testing phase) or by imposing sanctions on partners who refuse to deliver innovative solutions.

Network management

Whereas contract management is particularly focused on controlling the input (e.g., through innovation incentives in the tender award criteria) and output (e.g., through sanction management) of the process, network management is aimed at managing the collaboration process itself. Network management can be defined as 'the deliberate attempt to govern processes in networks' (Klijn et al., 2010, p. 1065). Literature on network governance and network management (e.g., Ansell and Gash, 2007; Emerson and Nabatchi, 2015; Huxham and Vangen, 2005; Klijn and Koppenjan, 2015; Klijn et al., 2010; Sørensen and Torfing, 2017) start from the premise that network complexities can only be solved by intervening in the collaboration process itself, instead of trying to control the input and output conditions (e.g., through contract management). Collaborative interactions are very dynamic, unpredictable, and interdependent, which means that the process needs to be controlled from the inside. By managing the interactions between the involved actors, network managers are able to align goals, vision, interests, and perspectives, and converge towards the same ideas. Network management resolves tensions and conflicts between collaboration partners and opens up discussions to explore the ideas and perspectives of the partners. As collaborative innovation is built on the collaborative interactions between the partners (i.e., synergy, learning, consensus building, commitment), innovation should also be stimulated by purposefully managing these collaborative interactions. Recent research has already indicated this stimulating effect of network management on collaborative innovation (e.g., Brogaard, 2017; Callens et al., 2021; Parrado and Reynaers, 2020).

Klijn et al. (2010) propose four different types of network management strategies. The first strategy is aimed at *exploring content*. This strategy searches for differences in perspectives and goals, which could strengthen the innovation process but might also hinder a smooth collaboration. Information and knowledge of partners is explored by using this strategy, and variation in ideas and solutions are identified by stimulating creative ideation. The second strategy involves *connecting* the partners together. The purpose of this strategy is to connect the perspectives, beliefs, and goals of the partners, but also their knowledge and resources. Klijn et al. (2010, p. 1069) point to several examples, such as the selective (de)activation of actors, resource mobilizing, the initiation of new series of interactions, coalition building, mediation, the appointment of process managers, and the removal of obstacles to and creation of incentives for cooperation. The third strategy is aimed at *arranging* the collaboration process. Whereas the governance structure represents the stable and rigid framework in which the collaboration process evolves, the collaboration arrangements that are stimulated through the arranging strategy are aimed at capturing ongoing collaborative interactions in new, ad hoc, and often temporary governance structures (e.g., boards, project organizations, etc.). The fourth strategy entails the establishment of *process agreements*. Process agreements refer to rules that the partners agree on in order to govern the collaboration process. Since a strong bureaucratic structure is rare in temporary partnerships, these rules allow the partners to clearly articulate what is expected from them and how the partnership will act in certain circumstances. Examples of such rules are rules for entrance into or exit from the partnership, conflict regulating rules, rules that specify the interests of actors or veto possibilities, rules that inform actors about the availability of information about decision-making moments, etc. (Klijn et al., 2010, p. 1069).

FEATURES OF THE INVOLVED INDIVIDUALS AND ORGANIZATIONS

Interpersonal and Organizational Trust

Systematic literature reviews on public service collaboration and innovation indicate that interpersonal trust is a crucial condition for enhancing the collaborative interactions between partners (Brogaard, 2021; Cinar et al., 2019; de Vries et al., 2015; Lopes and Farias, 2022; Voorberg et al., 2015). Trust can be defined as ‘a psychological state comprising the intention to accept vulnerability based upon the positive expectations of the intentions or behaviour of another’ (Rousseau et al., 1998, p. 395). The presence of trust between collaborating partners increases the confidence in the decisions and actions of the partners (McNamara, 2012) and reduces potential tensions and conflict between the partners (Entwistle and Martin, 2005). Interpersonal trust also facilitates the coordination and acceptance of the roles and responsibilities of the involved partners (Poocharoen and Ting, 2015). Increasing the trust between the involved individuals is necessary for processes of collaborative innovation, as the cultural diversity that is present in these processes might cause tensions and conflicts, which may reduce interpersonal trust (Diamond and Vangen, 2017). Similarly, because of the inherently risky nature of innovation, stimulating interpersonal trust will be of great importance in collaborative innovation processes (Brogaard, 2021). Moreover, creativity and innovation require a certain level of psychological safety, which allows individuals to freely think, and act without any hesitation, and which is fostered when people trust each other (Edmondson, 2003; Paulus and Dzindolet, 2008). Furthermore, interpersonal trust also eases collaborative interactions that are crucial for collaborative innovation, such as intensive engagement, discussion and dialogue, and commitment (Sørensen and Torfing, 2011; Torfing, 2019).

Knowledge and Skills

One of the principal reasons for establishing a partnership is the opportunity the collaboration creates to access desired resources (Ansell and Gash, 2007). Knowledge sharing through inter-organizational networks allows governments to make sense of a complex environment but also stimulates cross-fertilization of ideas from which innovations can arise (Hartley and Benington, 2006). The latter is clearly visible in Triple Helix configurations between industry, government, and university, in which different types of knowledge are united and feed into each other (Etzkowitz and Leydesdorff, 2000). Recent research of Trivellato et al. (2021) indicates that collaboration indeed allows knowledge sharing and learning (which are central to innovation), but also that these dynamics strengthen the innovation capabilities of the organization and the system. In other words, integrating the right knowledge pools can have profound effects on the capacity of the partnership, organization, and system to innovate. For this reason, it is useful to distinguish different types of knowledge.

Vines et al. (2015, p. 190) make a distinction between personal knowledge and explicit knowledge. Personal knowledge corresponds to subjective knowledge that is embodied in the individual talents, habits, and skills of people, and in the unconscious propensity of people to act in a certain way. According to Vines et al. (2015), this knowledge is often tacit (i.e., unconscious) and implicit (i.e., not yet made explicit), and is developed through experience. Explicit knowledge, however, is objective knowledge that is codified in a certain 'database' (e.g., in language). Whereas personal knowledge is experiential, explicit knowledge is often technical. Both are, however, needed in collaborative innovation processes, as the presence of explicit knowledge unites objective information from different fields of practice, while personal knowledge introduces intuition, know-how, and experience from these fields (Hartley and Benington, 2006). Selecting actors in the partnership who bring solid objective information to the table, while also having the experience and know-how of working in a particular field, will be particularly important in projects which aim at generating something new.

Skills for collaborative innovation, as a part of the personal knowledge of individuals, come in different forms. The two main activities in collaborative innovation (i.e., collaboration and innovation) each require a different set of skills. O'Leary et al. (2012) conclude from their empirical research into the skills of successful collaborators that there are three important groups of collaboration skills. The first group includes *individual attributes* such as, among others, having an open mind, patience, and self-confidence, and being risk-oriented, flexible, unselfish, persistent, and diligent. The second group includes *interpersonal skills* such as being a good communicator, an excellent listener, and being able to work well with people. The third group includes *group process skills* such as being able to facilitate, negotiate, solve problems collaboratively, deal with different personalities and organizational cultures, compromise, resolve conflicts, build consensus, and mediate. With regard to innovation skills, creativity and innovation literature indicates the importance of problem-solving skills (Lindsay et al., 2017) and creative-thinking skills (Anderson et al., 2014). Problem-solving skills refer to the ability to mobilize the proper personal and explicit knowledge for a certain problem (Vines et al., 2015). Creative-thinking skills correspond to a broad set of creative abilities, which essentially boil down to the core ability of divergent thinking, in which individuals refrain from drawing early conclusions, but consider multiple alternatives (Acar and Runco, 2012). Other authors point to the importance of having some previous experience with collaborative innovation (Sørensen and Torfing, 2017) and possessing the ability to critically question and evaluate assumptions (Sørensen and Torfing, 2021).

Besides collaboration and innovation skills, collaborative innovation processes are also highly dependent on who leads the process. Innovation leadership skills have been studied in many forms, from visionary leadership (van der Voet and Steijn, 2021), entrepreneurial leadership (Meijer, 2014), ambidextrous leadership (Giekse et al., 2020) in innovation research, to creative problem-solving leadership (Reiter-Palmon and Illies, 2004) and complexity leadership (Uhl-Bien et al., 2007) in creativity research. Similarly, collaborative leadership has also pointed to different leadership skills. For instance, Ansell and Gash (2012) propose a typology of three different leadership styles, each with its own skills (i.e., stewards, mediators, and catalysts). However, other authors have also looked at these (and other) leadership skills in collaborative innovation processes themselves. From this research, Lopes and Farias (2022, p. 124) extract leadership skills such as coordination capacity (Grotenberg and van Buuren, 2018), risk-taking (Mergel, 2018), commitment to the process organization (Hennala et al., 2011), and the ability to exert a certain patrimonial influence and authority over the process (Tuan, 2018).

External Support and Legitimacy

According to institutional theories of organizational development, organizational action is largely defined by the institutional environment in which the organizations operate. Through regulative, normative, and cognitive-cultural structures, the institutional environment gives meaning to the organization, and imbues the organization with legitimacy (Scott, 1995). In institutional theory, increasing organizational legitimacy is intricately connected with isomorphic pressures to conform to the value systems of the institutional environment (DiMaggio and Powell, 1983). The institutional environment defines what is appropriate for an organization and what is not. Institutional scholars propose that innovation is a strategy of the organization to conform better to the value system of the institutional environment (de Vries et al., 2015). For instance, Verhoest et al. (2007) show that state agencies that lack legitimacy are more likely to express innovative behaviour. However, successfully innovating to conform to the institutional environment is only possible if the organization receives signals from this environment, in the form of external support for the innovation. This external support may originate directly from the policy field in which the innovation is produced (e.g., the health sector for eHealth innovations) but might also come from other actors. For instance, failed public service innovations have been linked to a lack of support from political representatives and entities (Bakici et al., 2013; Cinar et al., 2019; Meijer, 2015), and media attention has been found to exert important pressures on innovation projects (Borins, 2001).

Partnerships and networks are particularly interesting to search for the effects of external support on innovation because they often connect multiple institutional environments with each other. Indeed, institutional scholars argue that institutional logics are being transmitted through these networks from one organization to the next (Owen-Smith and Powell, 2008). As these institutional logics can come from different institutional environments, different institutional environments might be responsible for imbuing the innovation with legitimacy. This might lead to value conflicts between the partners, as institutional environments of which some of the involved organizations are no part, can play an important role in legitimizing the innovation. As Klijn and Koppenjan (2015) argue, this institutional complexity can hinder productive collaboration. However, it might also affect how innovations are generated and eventually implemented. Even with a successful collaboration, innovations might still fail because they receive insufficient legitimacy from a particular institutional environment. Successful innovations might therefore need to integrate multiple institutional logics (e.g., public value and commercial logics).

USE OF INFORMATION AND COMMUNICATION TECHNOLOGY

Information and communication technology (ICT) has a central role in joining up governments, as they can structurally connect disparate entities through digital means (Dunleavy et al., 2005; Margetts and Dunleavy, 2013). As such, ICT works quite similarly to partnerships and networks, as it increases the capacity of organizations to share information, coordinate strategies and activities, and work closer together. Hence, the question is how collaborative innovation processes are stimulated through the use of ICT.

We consider two different ways in which the use of ICT can influence collaborative innovation. First, ICT can have an internal effect on collaborative innovation by enhancing the collaborative dynamics during the process. Indeed, digital technologies facilitate extensive communication and interaction between individuals, while reducing the transaction costs that such interactions would entail if digital means were absent. ICT provides opportunities to interact with each other more frequently and efficiently without the transaction costs that come with normal interactions (e.g., physically going to a meeting), and also enables communication with distant partners, which would be very time-consuming in other circumstances. Although intensive interaction is a necessary condition for collaborative innovation, it may, however, also cause 'collaborative inertia', which refers to slow, inefficient, and lengthy interactions, with a lot of deadlocks (usually caused due to tensions or conflicts), and which never really generate any action (Huxham, 2003). ICT might be better at directing certain interactions between individuals, as it allows quick, informal and bilateral communication (e.g., through digital messages such as email or message apps on smartphones). This could ensure that deadlocks and related tensions or conflicts, which hinder collaboration, are addressed more quickly. However, too much ICT-enabled interaction might also be a barrier to collaborative innovation, as building interpersonal trust and social identity can be more difficult without physical interactions.

Second, ICT can also have an external effect on the collaborative innovation process. Particularly if technological innovations are pursued by the partnership, broader ICT infrastructures will often play an important role in connecting the innovation to the digital systems of the involved organizations. These external ICT structures may not only influence the successfulness of the implementation of the innovation, but may also be crucial in the upscaling, diffusion, and broader adoption of the innovation by the public. For instance, Kattel et al. (2020) show how regional and national ICT networks often enable the creation and expansion of digital services, by allowing the connection of new services to the overarching ICT network. These ICT networks also frequently determine which organizations can work together, how the partners interact with each other, and which actions the partners can undertake (Kattel et al., 2020). Hence, the overarching ICT network and infrastructure might determine the starting conditions of these collaborative innovation processes (e.g., who is involved, which digital services are possible, which digital resources are available, etc.), which emphasizes the importance of this condition for collaborative innovation.

USER INVOLVEMENT

Services users are crucial stakeholders in collaborative innovation processes, and insufficiently involving them in innovation projects is regarded as a major barrier to public service innovation (Cinar et al., 2019). As users are the stakeholders that will apply the newly created service in practice, they are key to providing legitimacy to the innovation process and its outcome. When the partnership pursues legitimate and user-oriented solutions, this will require information about the expectations and demands of users, which will largely shape the development of the innovation (Sørensen and Torfing, 2017). A direct way for the partnership to obtain this information is by involving the users in the innovation process. Besides information regarding the expectations and demands of users, this also allows the partnership to access information that is extremely valuable but also difficult and costly to come by, such as information about service experiences and the local implementation context (Simmons and Brennan, 2017; von Hippel, 1994). Von Hippel (1994) calls this information 'sticky information', because it is difficult to acquire, transfer, and employ in a new context. Involving users in the innovation process makes (some of) this information accessible to the service providers. Roszkowska-Menkes (2017) discusses two additional reasons why users should be involved in innovation processes. On the one hand, users often have heterogeneous needs, which require a certain level of customization of services. Involving the users in the innovation process makes this customization easier. On the other hand, while users provide essential information for the innovation, they are usually not interested in shielding the innovation from competitors (e.g., by patenting the innovation) or commercializing the innovation. This expedites the broad diffusion and adoption of the innovation, which increases the potential implementation and impact rates of the innovation.

Literature on user involvement in processes of collaborative innovation relies particularly on coproduction research (Callens, 2022). Through coproduction, users can be actively involved in the collaborative innovation process. However, different types of user involvement can exist. For instance, users might be involved for different purposes (e.g., providing legitimacy or knowledge), in different stages of the innovation process (e.g., conceptualization stage or testing stage), in different intensities (e.g., isolated, ad hoc involvement or repeated, structural involvement), and in different ways (e.g., consultation, deliberation, development, etc.) (Alam, 2002). Furthermore, the coproduction process can be conditioned on the role of the user in the service system. For instance, the role of service planner and deliverer might be exclusively placed with the professional service providers, but users might also be partially or fully responsible for the service planning and delivery (Bovaird, 2007), which changes how these two stakeholders interact with each other in coproduction processes. Moreover, user involvement may also vary depending on the scope of the involvement. For instance, some users might only be involved on the operational level, while other users might be involved on the strategic level, or might even lead the whole process (i.e., respectively consumer, participative, and enhanced coproduction, Osborne and Strokosch, 2013). Alternatively, some user involvement might be aimed at service design *for* users (i.e., dominant role of service providers), *with* users (i.e., equal involvement of users and service providers), or *by* users (i.e., the dominant role of users) (Arnkil, 2010).

Multiple conditions have also been linked to successful (and unsuccessful) user involvement and coproduction, which can be clustered into two types of conditions. The first group of conditions relates to the capabilities of the users themselves. For instance, research has indicated that dialogue skills (Prahalad and Ramaswamy, 2004) and general psychological skills (Etgar, 2008) are important for successful user involvement. Additionally, Simmons and Birchall (2005) indicate that a certain degree of commitment to participate and invest time into the process is important for successful user involvement. Other scholars point to the qualities of the users that can inhibit effective user involvement. For instance, users might have a lack of motivation to radically innovate services or might have cognitive limitations (e.g., lack of knowledge) which hinders valuable input (Lettl, 2007). A second group of conditions corresponds to how the process of user involvement is organized and managed. For instance, a very rigid organization of the user involvement process, with specialized, isolated, and stable user tasks, might reduce the freedom for involved users to innovate (Torvinen and Ulkuniemi, 2016). Furthermore, managers should be aware to also involve 'unseen' users and to devote sufficient time to the active involvement of users (Gulliksen and Eriksson, 2006). Moreover, target groups for user involvement should be clear to the project manager, as should the proper balance between involving highly specialized users and involving users for legitimacy purposes (e.g., interest groups) (Karlsson et al., 2013). Also note that, although user demands remain essential to most user-oriented service innovations, some of the demands will be fixed and enforceable by, for instance, tenders, contracts, or grant agreements. This might cause tensions between what a public procurer or other financing institution demands, and what the users expect (Jæger, 2013).

CONCLUSION: TOWARDS A HOLISTIC APPROACH TO CONDITIONS OF COLLABORATIVE INNOVATION

This chapter provided an integrated framework for the conditions that influence processes of collaborative innovation for digital transformation. Through processes of synergy, learning, consensus building, and commitment (Ansell and Torfing, 2014; Sørensen and Torfing, 2011), innovation is stimulated in public-private innovation partnerships. Four clusters of conditions that work on these processes were identified, that is, conditions on the level of the partnerships, the involved individuals and organizations, the use of ICT, and the user involvement. These conditions may have isolated effects on collaborative innovation. For instance, by increasing the trust amongst the partners, collaborative interactions become smoother and more constructive, which enhances the various collaborative innovation processes. However, different conditions may also have a combined effect on collaborative innovation. For instance, different types of management techniques such as contract management and network management might have a larger effect on collaborative innovation when they are combined with each other.

Such a 'holistic approach' is particularly interesting for inherently complex and chaotic innovation processes, which are subject to interconnected dynamics that evolve simultaneously throughout the innovation process (Meijer, 2014). Hence, multiple conditions may influence the success of the collaborative innovation process at the same time. Employing such a holistic approach to the conditions of collaborative innovation might generate insights into the intricate nature of diverse, and sometimes even contradictory dynamics in the innovation process. For instance, Torfing (2019) mentions that collaborative innovation is inherently paradoxical as conditions that stimulate intense collaboration might also inhibit extraordinary innovation (e.g., collaboration thrives in contexts of similarity, while innovation exploits diversity). A holistic approach might uncover and explain such tensions and give accurate advice on how conditions of collaborative innovation lead to innovative public services.

Such a holistic approach on collaborative innovation is applied throughout Part III of the book, 'Public-Private Collaboration for Digital Transformation and Innovation'. In Chapter 8, we test the theorized conditions on a large empirical dataset of 19 eHealth collaborations in five European countries, through a qualitative comparative case study. Several of the identified conditions are then empirically tested in more detail in the subsequent chapters of Part III. Chapter 9 is devoted to the contract management and network management conditions, in which we assess whether the combination of these conditions stimulates the innovativeness of the produced eHealth services in these partnerships. Chapter 10 considers how users view their own roles in processes of user involvement. Finally, Chapter 11 tests how partnership design, and more specifically the structure of the social networks inside the partnership affects innovative outcomes.

NOTE

1. The TROPICO project received funding from the European Union's Horizon 2020 research and innovation programme under grant agreement No. 726840. This article reflects only the author's view and the Research Executive Agency of the European Commission is not responsible for any use that may be made of the information that the article contains. For more information: https://cordis.europa.eu/project/id/726840.

REFERENCES

Acar, S., and Runco, M.A. (2012). Creative abilities: Divergent thinking. In M. Mumford (ed.), *Handbook of Organizational Creativity* (115–40). Elsevier.

Agarwal, R., Gao, G.G., DesRoches, C., and Jha, A.K. (2010). The digital transformation of healthcare: Current status and the road ahead. *Information Systems Research, 21*(4), 796–809

Alam, I. (2002). An exploratory investigation of user involvement in new service development. *Journal of the Academy of Marketing Science, 30*(3), 250–61.

Anderson, N., Potočnik, K., and Zhou, J. (2014). Innovation and creativity in organizations: A state-of-the-science review, prospective commentary, and guiding framework. *Journal of Management, 40*(5), 1297–333

Ansell, C., and Gash, A. (2007). Collaborative governance in theory and practice. *Journal of Public Administration Research Theory, 18*(4), 543–71.

Ansell, C., and Gash, A. (2012). Stewards, mediators, and catalysts: Toward a model of collaborative leadership. *The Innovation Journal, 17*(1), 2.

Ansell, C., and Torfing, J. (2014). *Public Innovation through Collaboration and Design.* Routledge.

Appleyard, M., and Chesbrough, H.W. (2017). The dynamics of open strategy: From adoption to reversion. *Long Range Planning, 50*(3), 310–21.

Arnkil, R., Järvensivu, A., Koski, P., and Piirainen, T. (2010). *Exploring Quadruple Helix Outlining User-Oriented Innovation Models.* Final Report on Quadruple Helix Research for the CLIQ project.

Bakici, T., Almirall, E., and Wareham, J. (2013). The role of public open innovation intermediaries in local government and the public sector. *Technology Analysis & Strategic Management, 25*(3, SI), 311–27.

Bommert, B. (2010). Collaborative innovation in the public sector. *International Public Management Review, 11*(1), 15–33.

Borins, S. (2001). Public management innovation in economically advanced and developing countries. *International Review of Administrative Sciences, 67*(4), 715–31.

Bovaird, T. (2007). Beyond engagement and participation: User and community coproduction of public services. *Public Management Review, 67*(5), 846–60.

Brogaard, L. (2017). The impact of innovation training on successful outcomes in public–private partnerships. *Public Management Review, 19*(8), 1184–205.

Brogaard, L. (2021). Innovative outcomes in public-private innovation partnerships: A systematic review of empirical evidence and current challenges. *Public Management Review, 23*(1), 135–57.

Brown, L., and Osborne, S.P. (2013). Risk and innovation. *Public Management Review, 15*(2), 186–208.

Bryson, J.M., Crosby, B.C., and Stone, M.M. (2006). The design and implementation of cross-sector collaborations: Propositions from the literature. Special issue. *Public Administration Review, 66*, 44–55.

Bryson, J.M., Crosby, B.C., and Stone, M.M. (2015). Designing and implementing cross-sector collaborations: Needed and Cchallenging. *Public Administration Review, 75*(5), 647–63.

Callens, C. (2022). User involvement as a catalyst for collaborative public service innovation. *Journal of Public Administration Research and Theory, 33*, 329–41.

Callens, C., Verhoest, K., Garcia-Rayado, J., Van Doninck, D., Dockx, E., and Boon, J. (2020). *Practices of External Collaboration for Service Delivery – Research Report on the Conditions of Public-Private Collaborations to Foster Innovation*, https://tropico-project.eu/publications/.

Callens, C., Verhoest, K., and Boon, J. (2021). Combined effects of procurement and collaboration on innovation in public-private-partnerships: A qualitative comparative analysis of 24 infrastructure projects. *Public Management Review, 24*(6), 860–81.

Cinar, E., Trott. P., and Simms, C. (2019). A systematic review of barriers to public sector innovation process. *Public Management Review, 21*(2), 264–90.

Corsaro, D., Cantù, C., and Tunisini, A. (2012). Actors' heterogeneity in innovation networks. *Industrial Marketing Management, 41*(5), 780–9.

Crosby, B.C., 't Hart, P., and Torfing, J. (2017). Public value creation through collaborative innovation. *Public Management Review, 19*(5), 655–69.

Damanpour, F., and Schneider, M. (2008). Characteristics of innovation and innovation adoption in public organizations: Assessing the role of managers. *Journal of Public Administration Research and Theory, 19*(3), 495–522.

De Vries, H., Bekkers, V., and Tummers, L. (2015). Innovation in the public sector: A systematic review and future research agenda. *Public Administration, 94*(1), 146–66.

De Vries, H., Bekkers, V., and Tummers, L. (2018). The diffusion and adoption of public sector innovations: A meta-synthesis of the literature. *Perspectives on Public Management and Governance, 1*(3), 159–76.

Diamond, J., and Vangen, S. (2017). Coping with austerity: Innovation via collaboration or retreat to the known? *Public Money & Management, 37*(1), 47–54.

Di Meglio, G. (2013). The place of ServPPINs in the pange of public-private collaboration arrangements for services provision. In F. Gallouj, L. Rubalcaba, and P. Windrum (eds.), *Public-Private Innovation Networks in Services* (59–87). Edward Elgar.

DiMaggio, P.J. and Powell, W.W. (1983). The iron cage revisited: Institutional isomorphism and collective rationality in organizational fields. *American Sociological Review, 48*(2), 147–60.

Djellala, F., Gallouja, F., and Miles, I. (2013). Two decades of research on innovation in services: Which place for public services? *Structural Change and Economic Dynamics, 27*, 98–117.

Dunleavy, P., Margetts, H., Bastow, S., and Tinkler, J. (2005). New public management is dead – long live digital-era governance. *Journal of Public Administration Research and Theory, 16*(3), 467–94.

Earl, L. (2004). *A Historical Comparison of Technological Change, 1998–2000 and 2000–2002*. In the Private and Public Sectors, catalogue no. 88F0006XIE200407. Statistics Canada (Science, Innovation and Electronic Information Division). Ottawa.

Edler, J., and Georghiou, L. (2007). Public procurement and innovation – resurrecting the demand side. *Research Policy, 36*, 949–63.
Edmondson, A.C. (2003). Psychological safety, trust, and learning in organizations: A group-level lens. In R.M. Kramer and K.S. Cook (eds.), *Trust and Distrust in Organizations: Dilemmas and Approaches* (239–72). Russell Sage Foundation.
Edquist, C., Vonortas, N., and Zabala-Iturriagagoitia, J.M. (2015). Introduction. In C. Edquist, N.C. Vonortas, J.M. Zabala-Iturriagagoitia, and J. Edler (eds.), *Public Procurement for Innovation* (1–27). Edward Elgar.
Emerson, K., and Nabatchi, T. (2015). *Collaborative Governance Regimes*. Georgetown University Press.
Emerson, K., Nabatchi, T., and Balogh, S. (2011). An integrative framework for collaborative governance. *Journal of Public Administration Research and Theory, 22*(1), 1–29.
Entwistle, T., and Martin, S. (2005). From competition to collaboration in public service delivery: A new agenda for research. *Public Administration, 83*(1), 233–42.
Eom, S., and Lee, J. (2022). Digital government transformation in turbulent times: Responses, challenges, and future direction. *Government Information Quarterly, 39*(2), 1–9.
Etgar, M. (2008). A descriptive model of the consumer co-production process. *Journal of the Academy of Marketing Science, 36*, 97–108.
Etzkowitz, H., and Leydesdorff, J. (2000). The dynamics of innovation: From national systems and 'mMode 2' to a Triple Helix of university–industry–government relations. *Research Policy, 29*(2), 109–23.
Georghiou, L., Edler, J., Uyarra, E., and Yeow, J. (2014). Policy instruments for public procurement of innovation: Choice, design and assessment. *Technological Forecasting and Social Change, 86*, 1–12.
Gieske, H., Duijn, M., and van Buuren, A. (2020). Ambidextrous practices in public service organizations: Innovation and optimization tensions in Dutch water authorities. *Public Management Review, 22*(3), 341–63.
Grotenberg, S., and van Buuren, A. (2018). Realizing innovative public waterworks: Aligning administrative capacities in collaborative innovation processes. *Journal of Cleaner Production, 171*(10), S45–S55.
Gulliksen, J., and Eriksson, E. (2006). *Understanding and developing user involvement at a public authority*. Paper presented at the NordiCHI'06 Workshop on User Involvement and representation in e-Government Projects. Oslo.
Gullmark, P. (2021). Do all roads lead to innovativeness? A study of public sector organizations' innovation capabilities. *American Review of Public Administration*, 51(7), 1–17.
Hartley, J. (2005). Innovation in governance and public services: Past and present. *Public Money and Management, 25*(1), 27–34.
Hartley, J., and Benington, J. (2006). Copy and paste, or graft and transplant? Knowledge sharing through inter-organizational networks. *Public Money & Management, 26*(2), 101–8.
Hartley, J., and Rashman, L. (2018). Innovation and inter-organizational learning in the context of public service reform. *International Review of Administrative Sciences, 84*(2), 231–48.
Hartley, J., Sørensen, E., and Torfing, J. (2013). Collaborative innovation: A viable alternative to market competition and organizational entrepreneurship. *Public Administration Review, 73*(6), 821–30.

Hermans, F., van Apeldoorn, D., Stuiverd, M., and Kok, K. (2013). Niches and networks: Explaining network evolution through niche formation processes. *Research Policy, 42*(3), 613–23.

Hennala, L., Parjanen, S., and Uotila, T. (2011). Challenges of multi-actor involvement in the public sector front-end innovation processes. *European Journal of Innovation Management, 14*(3), 364–87.

Huxham, C. (2003). Theorizing collaboration practice. *Public Management Review, 5*(3), 401–23.

Huxham, C., and Vangen, S. (2005). *Managing to Collaborate: The Theory and Practice of Collaborative Advantage*. Routledge.

Innes, J.E., and Booher, D.E. (1999). Consensus building and complex adaptive systems. *Journal of the American Planning Association, 65*(4), 412–23.

Jæger, B. (2013). User involvement in public services innovation. In S.P. Osborne and L. Brown (eds.), *Handbook of Innovation in Public Services* (432–44). Edward Elgar.

Jobin, D. (2008). A transaction cost-based approach to partnership performance evaluation. *Evaluation, 14*(4), 437–65.

Karlsson, F., Holgersson, J., Söderström, E., and Hedström, K. (2013). Exploring user participation approaches in public e-service development. *Government Information Quarterly, 29*, 158–68.

Kattel, R., Lember, V., and Tõnurist, P. (2020). Collaborative innovation and human-machine networks. *Public Management Review, 22*(11), 1652–73.

Klijn, E.H., and Koppenjan, J. (2015). *Governance Networks in the Public Sector*. Routledge.

Klijn, E.H., Steijn, B., and Edelenbos, J. (2010). The impact of network management on outcomes in governance networks. *Public Administration, 88*(4), 1063–82.

Lam, W.F. and Li, W. (2018). Network structure and collaborative innovation processes – a comparative analysis of two elderly service networks in Shanghai. *Public Administration and Development, 38*, 87–99.

Lasker, R.D., Weiss, E.S., and Millier, R. (2001). Partnership synergy: A practical framework for studying and strengthening the collaborative advantage. *The Milbank Quarterly, 79*(2), 179–205.

Leiringer, R. (2006). Technological innovation in PPPs: Incentives, opportunities and actions. *Construction Management and Economics, 24*(3), 301–8.

Lettl, C. (2007). User involvement competence for radical innovation. *Journal of Engineering and Technology Management, 24*, 53–75.

Leydesdorff, L., and Meyer, M. (2003). The Triple Helix of university–industry–government relations. *Scientometrics, 58*(2), 191–203.

Lindsay, C., Findlay, P., McQuarrie, J., Bennie. M., Corcoran, E.D., and Van Der Meer, R. (2017). Collaborative innovation, new technologies, and work redesign. *Public Administration Review, 78*(2), 251–60.

Lindsay, C., Pearson, S., Batty, E., Cullen, A.M., and Eadson, W. (2020). Collaborative innovation in labour market inclusion. *Public Administration Review, 81*(5), 925–34.

Lopes, A.V., and Farias, J.S. (2022). How can governance support collaborative innovation in the public sector? A systematic review of the literature. *International Review of Administrative Sciences, 88*(1), 114–30.

Margetts, H., and Dunleavy, P. (2013). The second wave of digital-era governance: A quasiparadigm for government on the Web. *Philosophical Transactions of the Royal Society A Mathematical, Physical and Engineering Sciences,* 371, 1–17.

McNamara, M. (2012). Starting to untangle the web of cooperation, coordination, and collaboration: A framework for public managers. *International Journal of Public Administration, 35*(6), 389–401.

Meijer, A.J. (2014). From hero-innovators to distributed heroism: An in-depth analysis of the role of individuals in public sector innovation. *Public Management Review, 16*(2), 199–216.

Meijer, A. (2015). E-governance innovation: Barriers and strategies. *Government Information Quarterly, 32*(2), 198–206.

Mergel, I. (2018). Open innovation in the public sector: Drivers and barriers for the adoption of Challenge.gov. *Public Management Review, 20*(5), 726–45.

Mergel, I., Edelmann, N., and Haug, N. (2019). Defining digital transformation: Results from expert interviews. *Government Information Quarterly, 36*(4), https://doi.org/10.1016/j.giq.2019.06.002.

Nadkarni, S., and Prügl, R. (2021). Digital transformation: A review, synthesis and opportunities for future research. *Management Review Quarterly, 71*, 233–341.

O'Leary, R., Choi, Y., and Gerard, C. (2012). The skill set of the successful collaborator. *Public Administration Review, 72*(1), 70–83.

Osborne, S.P. (2006). The new public governance? *Public Management Review, 8*(3), 377–87.

Osborne, S.P., and Strokosch, K. (2013). It takes two to Tango? Understanding the co-production of public services by integrating the services management and public administration perspectives. *British Journal of Management, 24*, 31–47.

Owen-Smith, J., and Powell, W.W. (2008). Networks and institutions. In R. Greenwood, C. Oliver, K. Sahlin, and R. Suddaby (eds.), *The SAGE Handbook of Organizational Institutionalism* (596–623). Sage.

Parrado, S., and Reynaers, A. (2020). Agents never become stewards: Explaining the lack of innovation in public–private partnerships. *International Review of Administrative Sciences, 86*(3), 427–43.

Paulus, P.B., and Dzindolet, M.T. (2008). Social influence, creativity and innovation. *Social Influence, 3*, 228–47.

Paulus, P.B., Baruah, J., and Kenworthy, J.B. (2018). Enhancing collaborative ideation in organizations. *Frontiers in Psychology, 9*, https://doi.org/10.3389/fpsyg.2018.02024.

Picazo-Vela, S., Gutiérrez-Martínez, I., Duhamel, F., Luna, D.E., and Luna-Reyes, L.F. (2018). Value of inter-organizational collaboration in digital government projects. *Public Management Review, 20*(5), 691–708.

Poocharoen, O., and Ting, B. (2015). Collaboration, co-production, networks: Convergence of theories. *Public Management Review, 17*(4), 587–614.

Prahalad, C.K., and Ramaswamy. V. (2004). Co-creation experiences: The next practice in value creation. *Journal of Interactive Marketing, 18*(3), 5–14.

Provan, K., and Kenis, P. (2007). Modes of network governance: Structure, management and effectiveness. *Journal of Public Administration Research and Theory, 18*(2), 229–52.

Reiter-Palmon, R., and Illies, J.J. (2004). Leadership and creativity: Understanding leadership from a creative problem-solving perspective. *The Leadership Quarterly, 15*(1), 55–77.

Rogers, E.M. (2003). *Diffusion of Innovations*, 5th edn. Free Press.

Roszkowska-Menkes, M. (2017). User innovation: State of the Art and perspectives for future research. *Journal of Entrepreneurship, Management and Innovation*, *13*(2), 127–54.

Rousseau, D.M., Sitkin, S.B., Burt, R.S., and Camerer, C. (1998). Not so different after all: A cross-discipline view of trust. *Academy of Management Review, 23*(3), 393–404.

Schumpeter, J.A. (1942). *Capitalism, Socialism, and Democracy*. Harper.

Scott, W.R. (1995). *Institutions and Organizations*. Sage.

Simmons, R., and Birchall, J. (2005). A joined-up approach to user participation in public services: Strengthening the participation chain. *Social Policy and Administration, 39*, 260– 83.

Simmons, R., and Brennan, C. (2017). User voice and complaints as drivers of innovation in public services, *Public Management Review, 19*(8), 1085–104.

Sørensen, E., and Torfing, J. (2011). Enhancing collaborative innovation in the public sector. *Administration & Society, 43*(8), 842–68.

Sørensen, E., and Torfing, J. (2017). Metagoverning collaborative innovation in governance networks. *American Review of Public Administration, 47*(7), 826 –39.

Sørensen, E., and Torfing, J. (2020). Co-creating ambitious climate change mitigation goals: The Copenhagen experience. *Regulation & Governance, 16*(2), 572–87.

Sørensen, E., and Torfing, J. (2021). Radical and disruptive answers to downstream problems in collaborative governance? *Public Management Review, 23*(11), 1590–611.

Tadelis, S., and Bajari, P. (2006). Incentives and award procedures: Competitive tendering vs. negotiations in procurement. In D. Nicola, G. Piga, and G. Spagnola (eds.), *Handbook of Procurement* (121–39). Cambridge University Press.

Torfing, J. (2019). Collaborative innovation in the public sector: The argument. *Public Management Review, 21*(1), 1–11.

Torfing, J., Cristofoli, D., Gloor, P.A., Meijer, A.J., and Trivellato, B. (2020). Taming the snake in paradise: Combining institutional design and leadership to enhance collaborative innovation. *Policy and Society, 39*(4), 592–616.

Torvinen, H., and Ulkuniemi, P. (2016). End-user engagement within innovative public procurement practices: A case study on public–private partnership procurement. *Industrial Marketing Management, 58*, 58–68.

Trivellato, B., Martini, M., and Cavenago, D. (2021). How do organizational capabilities sustain continuous innovation in a public setting? *American Review of Public Administration, 51*(1), 57 –71.

Tuan, L.T. (2018). Behind the influence of job crafting on citizen value co-creation with the public organization: Joint effects of paternalistic leadership and public service motivation. *Public Management Review, 20*(10), 1533–61.

Uhl-Bien, M., Marion, R., and McKelvey, B. (2007). Complexity leadership theory: Shifting leadership from the industrial age to the knowledge era. *The Leadership Quarterly, 18*(4), 298–318.

Uyarra, E., Edler, J., Garcia-Estevez, J., Georghiou, L., and Yeow, J. (2014). Barriers to innovation through public procurement: A supplier perspective. *Technovation, 34*(10), 631–45.

Van der Panne, G., van Beers, C. and Kleinknecht, A. (2003). Success and failure in innovation: A literature review. *International Journal of Innovation Management, 7*, 309–38.

van der Voet, J., and Steijn, B. (2021). Team innovation through collaboration: How visionary leadership spurs innovation via team cohesion. *Public Management Review, 23*(9), 1275–94.

Verhoest, K., Verschuere, B., and Bouckaert, G. (2007). Pressure, legitimacy, and innovative behavior by public organizations. *Governance, 20*(3), 469–97.

Vial, G. (2019). Understanding digital transformation: A review and a research agenda. *Journal of Strategic Information Systems, 28*(2), 118–44.

Vines, R., Jones, M., and McCarthy, G. (2015). Collaborating across institutional and jurisdictional boundaries: Enabling the emergence of a national innovation system through public knowledge management. *Knowledge Management Research & Practice, 13*(2), 187–97.

Von Hippel, E. (1994). 'Sticky information' and the locus of problem solving: Implications for innovation. *Management Science, 40*(4), 429–39.

Voorberg, W.H., Bekkers, V.J.J.M., and Tummers, L.G. (2015). A systematic review of co-creation and co-production: Embarking on the social innovation journey. *Public Management Review, 17*(9), 1333–57.

Waldorff, S.B., Kristensen, L.S., and Ebbesen, B.V. (2014). The complexity of governance: Challenges for public sector innovation. In C. Ansell and J. Torfing (eds.), *Public Innovation through Collaboration and Design* (70–89). Routledge.

Walker, R.M. (2007). An empirical evaluation of innovation types and organizational and environmental characteristics: Towards a configuration framework. *Journal of Public Administration Research and Theory, 18*(4), 591–615.

Wilson, C., and Mergel, I. (2022). Overcoming barriers to digital government: Mapping the strategies of digital champions. *Government Information Quarterly, 39*(2), https://doi.org/10.1016/j.giq.2022.101681.

Windrum, P. (2008). Innovation and entrepreneurship in public services. In P. Windrum and P.M. Koch (eds.), *Innovation in Public Sector Services: Entrepreneurship, Creativity and Management* (3–20). Edward Elgar.

Wynen, J., Boon, J. Kleizen, B., and Verhoest, K. (2020). How multiple organizational changes shape managerial support for innovative work behavior: Evidence from the Australian public service. *Review of Public Personnel Administration, 40*(3), 491–515.

PART II

Intergovernmental collaboration for digital transformation

5. The coordination of digital government platforms: the role of administrative tradition and collaboration history

Jessica Breaugh and Steven Nõmmik

INTRODUCTION

Collaboration within the context of digitalisation processes represents opportunities to break down organisational silos that hamper bureaucratic public organisations, while at the same time offering a new way of approaching public service provision through streamlined services. It is also an essential component of public sector digitalisation approaches (Gil-García, 2012), both as a factor defining the change management process and a factor shaping the outcomes of this change given the new organisational structures that emerge from digitalisation processes (Kuipers et al., 2014; Orlikowski, 2007). In many cases, the process of digitalisation fundamentally changes the way both internal and external processes are conducted. With the introduction of digital platforms as well as the necessity for cross-governmental collaboration many challenges naturally arise. The challenges originate from both the siloed nature of administrative systems as well as the complex digitally specific context of increased power, complexity, and risk (Rackwitz et al., 2020).

Although technology implementation has been perceived as a technical exercise (for example, Kapoor et al., 2021), its introduction occurs in an institutional environment that considerably shapes its realisation (DeSanctis and Poole, 1994). This appears in goals set out in official strategic documents and the interpretation of these documents occurring from design to implementation. These variations emerge from varying cognitive frames of stakeholders who are embedded in different institutional environments (Orlikowski and Gash, 1994). With the increased role of government collaborative digitalisation projects, the potential challenges from the varying cognitive frames increase, with each actor bringing their own sets of goals and objectives, including protecting their own identities, processes, and approaches to service provision.

Despite this, however, research on collaboration in the context of government digitalisation is only just emerging (Costumato, 2021). Even fewer studies exist that examine intra-government collaboration from a comparative perspective (for example, Breaugh et al., 2023). A failure to compare collaborative approaches across different governments, and different types of projects, however, results in a lack of understanding of the role of context, or the forces external to digitalisation projects that play an essential role in how they develop and are ultimately implemented. According to Granovetter (1985), context is paramount to understanding behaviour and the outcomes.

Context can take on many meanings, however, as a starting point of comparative analysis two critical components emerge. First is the role of the *institutional environment*. Administrative tradition is useful for understanding the institutional environments that impact project dynamics and thus provides a method for studying the system context of projects (Breaugh and Hammerschmid, 2020; Painter and Guy Peters, 2010a). The second component is the role of *collaboration history*. Collaboration history in this context refers to the networks and relationships built in previous collaborations, which offer opportunities for knowledge sharing to initiate innovation (Damanpour and Schneider, 2006). This provides a method of understanding how relationships influence collaborative approaches. Therefore, the overarching question explored in this chapter is the influence of the institutional environment (understood through the concept of administrative traditions) and collaboration history (with a focus on pre-existing relationships) in the perceptions of power, complexity, and risk in digitalisation projects. This will be accomplished through an analysis of five case studies of collaborative digitalisation projects at the national level – all working on topics related to digital platforms linked to the European Union (EU) single digital gateway legislation.[1]

The organisation of the chapter is as follows. We will begin by outlining the concepts of power, complexity, and risk identified as common challenges that emerge in government digitalisation projects (Rackwitz et al., 2021). We then develop six propositions regarding how administrative traditions and collaborative history could be related to these challenges. We then present and discuss our analysis of the five case studies.

CHALLENGES OF DIGITAL COLLABORATION

While the uptake of collaborative solutions for large-scale government digitalisation projects is growing, three challenges are often highlighted in the literature that hamper and deter the collaborative process – particularly in a public sector context. These are power, complexity, and risk (Ansell and Gash, 2008; Crosby and Bryson, 2010; Osborne and Brown, 2011; Torfing, 2019). These will be explored in this section.

The first challenge to be discussed is power. Power is a Multi-faced concept reflecting the capacity to influence, control, or resist the activities of others (Purdy, 2012). The capacity derives from resources, formal structural positions, perceived importance as well as day-to-day activities (Choi and Robertson, 2014). While variations in power are inherent to multi-actor collaborations, its perceived misuse can create perceptions of power imbalances that hamper collaboration (Hartley et al., 2013; McGuire and Agranoff, 2011). Within collaborative initiatives, power can be used in different areas. These include deciding who is at the table, the design of the project itself, and what the governance structures used to manage it will be. The stakeholders included in the decision-making process affect the variation of perspectives available for defining the problem and choosing the solution (Ansell and Gash, 2008). Power imbalances may lead to the adoption of approaches from the more powerful members limiting a mutual understanding of the problem and the possible solutions (Bryson et al., 2006; McGuire and Agranoff, 2011; Wegrich, 2019). This can limit the capacity of less powerful members to contribute to the initiative (Jones and Hooper, 2017). More powerful members may also opt towards governance structures oriented in self-interest that ensure their resources and minimise risks (Wegrich, 2019). This is subject to constant shifts throughout the deliberative process, with the change in the relevance of individual stakeholders' goals, resources, and positions (Choi and Robertson, 2014). Within the context of digitalisation, power is open to further shifts concerning technological capacity itself. Technological capacity reflects the ability of an organisation to explore, develop, and/or adapt new technological solutions in public service design, delivery, and evaluation (Lember et al., 2018). Within a collaboration, actors with higher technological capacity are less dependent on collaborative functions, and thus yield considerable power in their levels of engagement and final adoption of the product.

The second challenge to discuss is complexity. Complexity is multi-dimensional, encompassing both the specific collaborative process as well as the wider administrative structure through substantive, strategic, and institutional facets (Klijn and Koppenjan, 2016). *Substantive complexity* refers to the complexity that arises from differences in perceptions of problems, goals, and foreseeable solutions within a collaboration project. This can be particularly acute in digitalisation projects encompassing many different service areas. The complexity surfaces both on the technical level with the interoperability of different digital solutions in a cross-organisational context and the semantic level with a shared understanding of work language (Gil-García and Sayogo, 2016; Pardo, 2010; Picazo-Vela et al., 2018). Each collaboration partner is likely to have different goals for a particular digital product, and thus propose very different solutions. *Strategic complexity* refers to complexity related to varying strategies used by actors within a collaboration to handle conflicts. Existing organisational goals and values guide the actors towards behaviour that leads them towards a desirable goal (Klijn and Koppenjan, 2016; Raadschelders and Whetsell, 2018). These values originate from personal beliefs, organisational, and professional culture as well as society guiding an individual participant towards a specific choice (Raadschelders and Whetsell, 2018). Varying strategies can lead to uncoordinated actions, further diverging the project outcomes from their intended goals. Finally, *institutional complexity* reflects the institutional context in which a collaborative project is embedded, including formal and informal traditions, laws, and regulations.

The third and final challenge for collaboration is risk. Public sector organisations have a reputation for being risk averse. Risk aversion comes from several factors – from a lack of resources to a very limited tolerance for failure (Flemig et al., 2016). Furthermore, risk perception is contingent on the organisation's previous history of reforms and the capacities (including technological capability) developed during reform phases that affect the ability to evaluate risk with an initiative (Kattel et al., 2020; Torugsa and Arundel, 2017). From a single-actor perspective, engaging in a collaborative project opens the risk of losing control of their autonomy, a loss of legitimacy and resources, or creating the fear of blame for potential failures (Hinterleitner and Sager, 2015). It may also result in a lack of engagement from critical actors. Within the context of digitalisation projects, an elevated level of uncertainty due to the high financial costs, accountability, and knowledge asymmetries elevate these risk perceptions (Mergel, 2016; Neumann et al., 2019). Therefore, the challenge becomes how to manage risk perceptions, rather than the risk itself (Timeus, 2018; Timeus and Breaugh, 2020). Having explained three core challenges related to collaboration within a digital context, the discussion will now shift to understanding how contextual factors of administrative tradition and collaboration history are linked to these challenges. These will be explored in the next section.

SYSTEM CONTEXT AND COLLABORATION CHALLENGES

Two aspects of system context relevant to digitalisation are the *institutional environment* of the projects, defined using the concept of administrative tradition, and *collaboration history*, defined as the pre-existing relationships apparent at the onset of a collaborative project. In this section, we develop argumentation regarding how these aspects play a role in understanding the differences in the challenges (i.e., power, complexity, and risk) present in collaborative digitalisation projects.

Administrative Tradition

Although the design and implementation phase of digitalisation initiatives possess an important role in shaping the final adoption and diffusion, the initial steering originates from the facilitating and constraining role of the underlying structures within the environment (Barrett et al., 2013). This is concluded within both the broader public sector innovation literature (for example, de Vries et al., 2016) and the digital innovation literature (for example, Nasi et al., 2015). With considerable perceived differences present in existing administrative systems, administrative traditions provide a good heuristic tool for detailing this context. Administrative traditions refer to a way of classifying public administrative systems based on social and political institutions vis-à-vis society, the history of a particular system as well as the legal and regulatory systems (Painter and Guy Peters, 2010b). Research in this area has suggested that traditions can and do explain differences in state behaviours, including the legal and/or cultural constraints of individual government actors when they collaborate with one another. Depending on the administrative tradition present, the stakeholders may be predisposed to several values related to efficiency, effectiveness, and quality (Guy Peters, 2021). Anglo-Saxon traditions tend to be characterised as non-legalistic, pragmatic, and pluralist, with a strong centralist governance structure (Huxley et al., 2016; Painter and Guy Peters, 2010a). The Continental tradition is characterised by a strong legal basis for governance and as interventionist in their approach to society as well as a strong hierarchical governance approach (Meyer and Hammerschmid, 2010). They also have a clear separation of federal and regional powers, characterised by a strong adherence to the principle of subsidiarity (Benz and Zimmerman, 2011). Finally, the Scandinavian traditions appear to be a mix of both Anglo-Saxon and continental approaches. On the one hand side, they have a strong legal tradition as a basis for understanding the state and a consensus-oriented approach (Huxley et al., 2016), but on the other hand side, they mirror the centralist structures of the Anglo-Saxon models with a stronger corporatist structure (Meyer and Hammerschmid, 2010). Based on these characterisations, we would expect to see differences in how the projects are designed and managed that reflect, in part, administrative and cultural traditions in the given country. We propose six propositions to aid in this analysis.

With the centralised governance structures present in Anglo-Saxon countries, there is a larger tendency towards asymmetrical power balances with the leading organisation(s) controlling stakeholder activation (Huxley et al., 2016). This is compared to the Continental and Scandinavian contexts, where power is more multi-lateral with more limited shifts to alternative power positions due to a stronger adherence to established processes (Bach et al., 2017). With stronger asymmetries present, there may be a higher tendency towards negative responses due to perceived power imbalance that can lead to a reduction in commitment and turf protection strategies from actors who are perceived to have a limited voice. Proposition one (P1) therefore argues that there will be a stronger perceived power imbalance within the Anglo-Saxon countries compared to Continental and Scandinavian traditions.

Our second proposition (P2) focuses on the concept of complexity. As actors face differences in understanding technical solutions as well as semantics from professional language, complexity can increase. The administrative structure within the Continental tradition focuses on a balance between procedures and rules, maintained through public officials, trained typically as lawyers, within the policy fields, leading to challenges in the initiation of boundary-spanning initiatives (Pollitt and Bouckaert, 2011). In some cases, complexity may be heightened due to the necessity for legal frameworks to be developed alongside the projects in the Continental countries, while the pragmatism of the Scandinavian and Anglo-Saxon traditions could limit the plurality of perspectives in favour of cost-efficiency, therein reducing the substantive and strategic complexity. We, therefore, propose (P2) more decentralised and shared governance structures, for example, seen in the Continental tradition, can increase substantive and strategic complexity.

Our third proposition (P3) focuses on risk. A general aversion to risk in the public sector is quite well established in the literature (Chen and Bozeman, 2012). However, scholars have noted the role that socio-political institutions play in risk perception (Slovic, 1997). We, therefore, suggest that administrative traditions with more ministerial independence and pragmatist approach, like the Anglo-Saxon tradition, may experience more risk perceptions with stakeholders being more risk averse to participate in collaborative initiatives. The combination of ministerial independence, lack of legally binding contracts, and pragmatism lead stakeholders to an increased perception of the possible losses to other stakeholders from failure, thus affecting their calculations. With the silosation present, the mitigating factors from the past collaboration are limited due to limited connections. On the other hand, the continental European traditions, characterised by a more strict socio-legal order may perceive digitalisation projects as riskier in general as many disrupt the stability of a legal state – often pushing for reforms and change at a faster speed than the administration can process.

Collaboration History

Next to the institutional perspective of administrative tradition, the collaboration history of stakeholders provides additional contextual factors facilitating or hindering collaborative initiatives. Based on previous research, it is evident that relationship building is a critical component in collaboration projects and is linked to the performance of the collaboration in general (Chen et al., 2019; Luna-Reyes et al., 2007). With an increased level of social capital, project norms and behaviours develop much faster, increasing the probability of developing clear converging goals, as well as active and effective decision-making strategies (Fedorowicz et al., 2014). This provides stakeholders access to a more informal environment, where formal rules are of lesser importance, thus facilitating more knowledge sharing (Lewis et al., 2018). Collaborations that start with successful collaboration history are more readily able to function due to the activation of social capital (Granovetter, 1973). Successful collaboration history facilitates the pre-eminence of personal relationships over formal role relationships, which affect the evaluation of benefits and the assignment of obligations for different stakeholders (Ring and de Ven, 1994). This is underscored by Dawes and Pardo (2002), who note that collaboration history impacts the way new projects emerge and develop. In the context of digital collaborations, previous interactions may result in a better understanding of the digital infrastructure of potential partners as well as in initial compatibility between the solutions (Kattel et al., 2020). If the past was positive, it makes relationship building easier as there is already a base level of trust and understanding.

Past relationships may increase the acceptance of created asymmetries with stakeholders improving mutual understanding and actions to reduce the distance (McGuire and Agranoff, 2011). With the social capital present, the possible negative perceptions of power imbalances are appeased. Within digital initiatives, stakeholders have adopted a common language and better comprehend the underlying technical infrastructure stakeholders possess providing a better idea of the possible solutions (Pardo et al., 2010). Therefore, our fourth proposition (P4) argues that past successful collaboration history reduces negative perceptions of power imbalance.

With the increased mutual understanding from previous collaborations, it is easier for stakeholders to comprehend the technological capacity and goals of stakeholders (Chen and Lee, 2018). Through a common language, the stakeholders are better able to perceive the differences in interests, resources, and competencies (Quick and Feldman, 2014). This streamlines the process that stakeholders would otherwise utilise for getting to know each other. Therefore, projects that have actors with a positive collaborative history perceive substantive issues more compared to stakeholders without a collaborative history. With a lack of past collaborative history, the stakeholders assume a position based on their existing role within the administrative structure and are more rigid towards adjustments (Alford and O'Flynn, 2012). The appeals to existing strong relations result in the stakeholders perceiving strategic and institutional issues less acutely. Therefore, our fifth proposition (P5) is that collaborations with positive past relationships predominantly perceive substantive complexity, whereas stakeholders lacking past relationships perceive strategic and institutional complexity more strongly.

Moving to the concept of risk, high levels of trust and the predictability of other stakeholders' actions lead other partners to be more confident in committing resources and more flexible towards any potential shifts in role requirements from the initial agreed setup, thus reducing the effect of challenges (Alford and O'Flynn, 2012; Edelenbos and Klijn, 2007). Past positive relationships improve the feeling of continuity and provide a shared understanding of the group and therefore limit the perception of risk (Joffe, 2003; Noteboom et al., 1997). If the past was negative, however, it could make the starting conditions more complex and induce higher risk perceptions. The lack of past interactions results in stakeholders being more cautious when engaging with stakeholders (Bryson et al., 2006). Negative interactions limit the willingness of actors to engage in trust-based agreements and make them rely more on existing institutional design (Nooteboom et al., 1997). Based on past failures, stakeholders perceive a higher risk of potential resource loss (Verhoest et al., 2007). Therefore, our sixth proposition (P6) is that past positive relationships would reduce the negative perceptions of risk.

CASE STUDY INTRODUCTION

The cases included are key national projects within the EU's Single Digital Gateway (SDG). The SDG focuses on the digitisation of public services across the EU. They all represented key administrative traditions in Europe (based on the classification from Meyer and Hammerschmid, 2010). The SDG not only provided specific targets but also outlined principles in which this should be undertaken. These include the once-only principle and being user-centric. Because most citizen services cross organisational boundaries – especially to achieve the once-only principle – collaboration is at the forefront in terms of how projects are designed, built, and ultimately implemented. A total of 36 interviews across the five cases inform the empirical analysis of this chapter. There were between six and nine interviews per case, where at least one senior manager, consultant, and programme manager were interviewed. For more information about the cases, see Breaugh et al. (2023). All interviews were transcribed and coded (by coders in their original language). Each interview was coded by one person and reviewed by a second. The coding scheme was developed by a main coding team, and several training sessions were held for the coders for each case study. The coding scheme was based on a combination of concept-driven and data-driven approaches. Coders were instructed to highlight segments of interviews when the interviewee discussed aspects related to risk, complexity, and power imbalances based on a clear coding book and framework based on Rackwitz et al. (2020). The interview questionnaire

Table 5.1 *Case characteristics*

Country	Belgium	Denmark	Estonia	Germany	United Kingdom
Administrative Tradition	**Continental**	**Scandinavian**	**Scandinavian**	**Continental**	**Anglo-Saxon**
Project	**Civil Registry**	**eIDas Regulation**	**Employment Registry**	**Online Access Act (OAA)**	**Government as a Platform (GaaP)**
Project duration	2010–19	2013–ongoing	2013–ongoing	2017–22	2011–ongoing
Main goal	Moving the decentralised Civil Registry to a central online database/ mandatory	Implementing electronic identity verification/ voluntary	Creating a centralised digital solution for the collection and storage of all employment data	Offering all public services online via a joint portal/ mandatory	Introducing GaaP as an overarching platform principle/ voluntary
Key Lead	Administrative Simplification Service	Agency for Digitalisation	Estonian Tax and Customs Board (ETCB)	BMI/it-PC	Government Digital Services (GDS)

Source: Adapted from Breaugh et al. (2020).

included questions about power, complexity, and risk, enabling the authors to identify the relevant themes in the interview data. Periodic quality checks, including different coding exercises, were undertaken to encourage consistency among the coders (for more information regarding the coding process, see Breaugh et al., 2023). Table 5.1 presents the characteristics of each of the cases.

FINDINGS

Administrative Tradition and Collaboration Challenges

Overall, it was clear that the institutional environment played a role in if and how decisions were made. Exploring the proposition of power imbalances (P1), interviewees from all the cases highlighted perceiving power imbalances. However, based on the coded data, this was perceived most acutely within the UK case, where the context had a contributing role to the perception of challenges. The key project coordinator, the Government Digital Service

(GDS), faced obstacles related to the voluntary nature of the project. With the surrounding context of strong ministerial prerogative (a clear attribute of the Anglo-Saxon tradition), budgetary framing, and silosation, the stakeholders perceived the initiative as a shift in status quo towards GDS obtaining a more asymmetrical position at the expense of their resources that led to confrontations.

> … GDS, when it first started, grew up in opposition to the other departments because, essentially, building GOV.UK took capabilities that other departments had, and centralised them outside of those departments. (UK SDG 2, Pos. 36)

The asymmetrical position formed was related to challenges within their administrative tradition – with the necessity to break down hierarchical structures within the ministries. This led to cross-organisational engagement being affected by the existent structure.

> The truth is that there are some problems with collaborating with departments if you think about them in terms of departments because departments are run in a top-down way by people that in the civil service often have highly politicised careers where they're trying to get promoted. They have their own power struggles. So, actually dealing with departments from the top down is, I think, a really difficult and problematic way to transform government, because the people at those top levels don't actually … They're so far removed because of the way the civil service is structured in terms of its hierarchy. They're so far removed from the needs of the people doing the work of delivering services at the front line that they make bad decisions for them. (UK SDG 2, Pos. 43–44)

While there were power struggles present with the other cases and traditions, they weren't perceived as the most important challenge. Within the Continental traditions, the context provided measures for mitigating perceived challenges. The German and Belgium cases were legally mandated which shaped the dynamics from power imbalance to power management because stakeholders *had* to engage with the process and it was clear from the onset, through legal mandates, the roles and responsibilities of the actors. This limited the opportunities to exercise power for the creation of asymmetries. Furthermore, the federal structures present in both Belgium and Germany also meant that power was more dispersed between the partners (regardless of size), which meant that leaders needed to engage in more negotiation and mediation. In the German case, for example, stakeholders even began to align with the departments' objectives to be able to receive support.

> In the past, there were always laws, I would say from the Federal Ministry of the Interior for the internal issues. Then there were laws, you can think of the examples at will, quasi-always laws from the corresponding ministry for this specialist

> context, but very rarely only cross-cutting. And the OZG is something cross-cutting, which also pushes us in particular in our federalism. (DE SDG 2, Pos. 29)

With the Estonian and Danish cases, the coordinators possessed a considerable number of resources and an asymmetrical position like the UK case. This was manifested in the control of the interaction arenas (regarding stakeholder activation as well as the agenda) and resources (funding, personnel, knowledge). However, the engaged stakeholders perceived limited challenges from power imbalances. The context of both countries possessed a role, as informality, high levels of trust, merit-based system, and low levels of politicisation led to both countries framing the initiative from a technical perspective.

To explore proposition 2 (P2), we examined the impact of administrative traditions on complexity. The German and Belgium cases, representing the Continental tradition, highlighted the importance of the legal basis for providing the necessary authority and the frame of reference for enacting the change. This limited substantive complexity. Both cases, however, noted a challenge in strategic complexity through issues of interoperability of the digital architecture between the individual ministries. In the German case, this was the result of the heterogeneity of the existing functionalities of the actors and the challenges of funnelling them into a joint portal. In the Belgium case, heterogeneity was also present through the varying capacities and IT systems present in local municipalities. However, due to the digitalisation of a previously physical process, the decentralised digital infrastructure possessed a limited role, with the strategic complexity originating more from the deep-seated identities and silos of government ministries.

With the Scandinavian tradition, through the Estonian and Danish cases, there were certain similarities as well. The focus on limited politicisation, professionalisation, pragmatism, and informal environment streamlined the process, limiting the perceptions of complexities to a primarily substantive level. The substantive complexity of the initiatives was perceived in both cases with differences in operational logic leading to miscommunication regarding functionalities and opportunities. Both initiatives highlighted the importance of perceiving the process as a technical challenge rather than a political one. This was achieved through the framing of the technical capabilities of the coordinators, with both the Danish Agency for Digitisation and the Estonian Tax and Customs Board utilising similar measures (i.e., existing resources and past experiences). The engagement within the Estonian and Danish cases highlighted some elements of corporatism. The Danish case highlighted attempts to engage as many public sector actors as possible with the scope and design of the Gateway with a strong mutual understanding regarding the possible efficiency wins through the initiative. Within the Estonian case, through mutual recognition of the problem and the administrative burden, the stakeholders attempted to engage not only the public but also interest groups within the private sector to build mutual understanding regarding steps forward. Informal networks were used between middle and top-level managers of the respective agencies with the initiative reaching the political level. The broad level of engagement contributed to the substantive complexity with stakeholders possessing challenges even with the use of proper professional language. For example, in Estonia, one interviewee noted,

> The main differences were with regards to the level of detail in the registry entries and the classification to be created next to it. Whether the dataset that was being entered was sufficient and mutually understood. For instance, there was even a very large debate with regard to employment contracts and the use of the term contractor, which is completely unacceptable to us. As the employment contract can be only between an employee and an employer with contractors being covered through the Law of Obligations Act. … Devil is in the details. These tiny details would sometimes make us argue for hours and reaching a common understanding in these instances was difficult. (EE SDG 3, Pos. 103)

In the UK case, the connections between administrative traditions and complexity were less pronounced. The interviewees highlighted the challenges with complexity originating from the complicated UK administrative structure with the taxonomy of services delivered and the people responsible obfuscated. In conjunction, a rotating of leaders and personnel (also a typical characteristic of Anglo-Saxon states) meant that the complexity of the project was strategic – including convincing individual ministries to become engaged in the project (thus giving up their resources). These issues started right from the beginning, with the GDS's struggles with identifying key stakeholders.

> I think you need to have all the people that matter in the room at the start. Very often we would start with the first person we found, which was because we were under time pressure, which is the wrong way to go about it. You need to bring people together, get all the people in the room that need to be there to start with, to start together. Starting together is really important for collaboration. (UK SDG 2, Pos. 165)

However, hierarchical governance was present through the GDS, who used their asymmetrical position to determine the products to be developed as well as the stakeholders to be involved in new initiatives.

With proposition three (P3), the connections between administrative traditions and risks remained ambiguous. The Estonian and Danish cases highlighted the role of corporatism and the consensus-oriented approach. While the Estonian and Danish cases relied on pre-existing positive relations, the interviewees perceived important risks at the different stages of the initiatives. Within the Estonian case, the engagement of different stakeholders led to an increased perception of risks regarding efficient service provision due to differences in operational logic and standards.

> If insurance is missing, they will call us [Estonian Health Insurance Fund]. If they start checking an individual for health insurance that should already exist but hasn't been established due to technical reasons, then the individual has to pay for their health care service on their own, which they shouldn't need to do. (EE SDG 5, Pos. 59)

The risks regarding effectiveness and compatibility between processes and digital solutions were also reflected in the Danish case. This was highlighted by the aim of the stakeholders (i.e., Tax Agency) to keep digital development to a minimum to avoid either duplication or conflict with existing information systems. Furthermore, the consensus-building environment amplified perceptions of risks from time pressure for the Estonian case, as there were disagreements regarding the legal framework between actors (i.e., Ministry of Justice and Tax and Customs Board).

In the UK case, the perceived risks had similarities with the Danish and Estonian case reflecting the similarities between the respective traditions. The UK case reflected risks regarding effectiveness with stakeholders concerned about the productiveness and sustainability of the technical solutions for which individual ministries would remain responsible. The pragmatist and centralist-oriented structure created further perceptions of risk. For interviewees, this manifested with the perceived risk of GDS's reputation and the ability to manage expectations and narratives with issues of overpromising and balancing between meeting desired goals and implementing the solution promptly.

> The story of GOV.UK Verify is one where the concerns about risks of the viability of the product, on the side of customers in government, government service teams, departmental leads, whoever it is, in terms of those actors, has meant that there is a narrative around the product. Which is that it is troubled, it doesn't work, it's a problem, etc., etc. And partly that's a symptom of it being such a high-profile thing. (UK SDG 1, Pos. 121)

Many of the organisations seeking digital solutions saw the processes as risky because they did not have long-term funding for the maintenance of the programmes, especially coming from outside their ministerial prerogative.

In the Belgium and German cases, both cases perceived risks of failure and loss of resources and authority. With the pressure to standardise, the decentralised structure provided a space for increased perceptions of risks with uncertainty regarding the resource allocation moving forward. In the German case, the risk aversion behaviour entrenched within the administrative culture became a considerable obstacle. This stemmed from attempts to maintain the resources available to organisations due to the perceived risk of being transferred through digitalisation as well as the reputational damage due to the potential failure. While Belgium also had perceived risks linked to the decentralised structure and the dispersion of resources, the bigger risks lay in the resistance by the Ministry of Internal Affairs and the Ministry of Justice.

Overall, the first two propositions concerning administrative traditions and power and complexity (P1, P2) received support with the cases highlighting the impact of the unique factors within different administrative traditions related to the digitalisation initiative. The silosation and decentralisation of competencies present in the Continental tradition resulted in power games as well as perceptions of strategic complexity through goal conflicts with stakeholders. This was contrasted by the Scandinavian tradition, where a combination of professionalisation, pragmatism, and informality led to stakeholders viewing the process through a lens of technical issues rather than political ones, leading to challenges of substantive complexity. The final proposition regarding administrative traditions and risk (P3) receives limited support. Although all of the cases reported risk, the sources of this perceived risk differed (perceived loss of resource, reputation, power, etc.). For example, the Scandinavian perceptions derived from compatibility while the Continental tradition derived more strongly from turf protection, and the Anglo-Saxon tradition risks were more concerned with ministerial independence. It was clear that during a digitalisation process, the administrative tradition can amplify potential collaborative challenges, but also mitigate them or provide context-specific intervention measures.

Collaboration History and Collaboration Challenges

The second research question examines the role of collaboration history. Propositions four to six (P4–P6) proposed that positive collaboration history would influence how power, complexity, and risk are perceived in the projects. To explore proposition four (P4), we focus on the links between power imbalances and collaboration history. Within the Estonian case, the interviewees emphasised the considerable role of collaboration history in accepting power imbalances. For example, despite stakeholders possessing asymmetric positions within the initiative (with available resources as well as organisational priorities), the stakeholders were accepting of these disparities. The interviewees noted several interconnected factors (i.e., the technical capability of the coordinator, past successful initiatives, access to increased resources, and trust of individuals connected to the initiative) with some of the factors being based on interdependencies from previous collaborations:

> As I said before, the reason why the ETCB came to us, was because of our very good working relationship. We used a lot of their data in certain activities. The use of the data also meant that the system was interfaced. We had already developed a good working relationship and I would also point that out as far as the ministries were concerned. (EE SDG 3, Pos 9)

This was utilised by the coordinator to frame the project as a technical rather than political project despite the shifts in power positions following implementation. The Denmark case utilised a similar framing with power imbalances present, but not perceived as a significant challenge. However, the interviewees did not highlight the role of collaboration history, but rather the technical capability and past successful initiatives of the coordinator as key. The Belgium case also utilised technical capability to reduce perceptions of power imbalances with a weaker past collaboration history with the stakeholders. However, a lack of collaboration history meant that the project leaders had to spend more time and effort to establish themselves and their own perceived legitimacy. This was also like the UK case, in that the legitimacy of the project leaders needed to be established. The German case reflected an antithetical case with considerable fragmentation and heterogeneity resulting in clear power imbalances that led to impediments with active participation by the stakeholders with a stronger power position. The aforementioned examples highlight the impact of collaboration history on power imbalances. It is clear that collaboration history has ameliorating effects in accepting the power imbalances of the collaboration projects, while the administrative traditions have a closer link to understanding how power imbalances emerge in the first place.

Proposition 5 (P5) focused on the connections between collaboration history and complexity. The German case highlighted complexity as the most perceived challenge for the initiative. However, while the fragmented institutional environment inhibited considerable pre-existing collaboration ties, the interviewees did not highlight weak collaboration ties as a factor. In the UK case, with a lack of strong collaboration history, the coordinator was initially unaware of the additional resource required for mapping the necessary stakeholders and services due to institutional complexity.

> In some cases, quite high-profile services ... we went round in circles, cats from team to team, to team, to team. In the end, we couldn't find someone, who was responsible for it, so it wasn't included in the research. ... it wasn't always easy to find the actual responsible party for it. (UK SDG 1, Pos. 38)

The Belgium case also had a limited collaboration history, and they also experienced significant institutional complexity. This resulted from the varying capacities of the municipalities as well as the legislative framework that needed to be complied with. The interviewees for the Estonian case didn't perceive complexity as a prominent challenge with past collaboration highlighted as a reason for only minor substantive complexity being present. Within the Danish case, substantive complexity was also highlighted as a minor challenge. This was mainly to do with user definition and subsequent development. However, the Danish case had a limited collaboration history between the stakeholders. While collaboration history can have ameliorating effects regarding the overall perception of the challenges from complexities present, the administrative traditions help to comprehend the type of complexities the stakeholders perceive as most acute.

Proposition six (P6) focuses on the link between collaboration history and risk. For the Estonian case, there was very limited risk perceived, and it was clear that past successful collaboration history was a reason for this. The interviewees from the Belgium case were more positive with risks being of limited importance. The main risks were the tensions between the National Register and Civil Registry and between the Ministry of Internal Affairs and the Ministry of Justice. For the German case, the fragmentation was reflected in the perceived risks regarding technical challenges with the creation of a joint portal in the context of heterogeneity. The interviewees also noted that there was an intra-organisational risk averse culture that affected decision-making. A lack of previous collaborative experience meant that in both cases, the senior leaders had to spend time and effort to solve these challenges and build their legitimacy. In Denmark, the collaboration history did not appear to impact risk perceptions per se as this was overshadowed by technical issues. In the UK case, the interviewees perceived several risks. The lack of collaboration history was reflected in uncertainty regarding the capabilities and achievability of goals and overpromising issues. Overall, we appear to have only limited support for collaboration history and risk perceptions, following similar findings with administrative tradition.

Overall, the analysis of connections between collaboration history and challenges (P4–P6) highlighted certain tendencies. The analysis provided the strongest evidence for P4, with the existence of strong collaboration history enabling stakeholders to utilise the established relationships for idea creation and conflict resolution to streamline interactions between stakeholders. Within the Estonian case, it was an essential component for the success of the initiative. For the UK, the lack of strong collaboration history resulted in GDS needing to put considerable effort into engaging stakeholders and determining the relevant stakeholders from both the top and bottom of the hierarchies. With regards to P5, while collaboration history was reflected by the stakeholders related to complexity in the Estonian and Belgium cases, there was only limited support for the propositions. Although the Estonian case highlighted substantive complexity and its limited challenge, the Danish case highlighted perceiving similar challenges with limited collaboration history, making it difficult to make any conclusions regarding the connections between collaboration history and complexity. For P6, the findings indicate that the previous collaboration history ameliorates the perceived possible risks, but the evidence is limited and not consistent across the cases.

CONCLUSION

The purpose of this chapter was to examine the collaborative dynamics of, power, complexity, and risk in large-scale digitalisation projects across different administrative traditions with different histories of collaboration. To do so, we add to the literature that examines cases in the context of institutional embeddedness. We proposed six propositions, organised between the themes of administrative tradition and collaboration history and their link to common collaborative challenges. The findings show full support for P1, P2, and P4 and limited support for P3 and P5. This means that there is evidence to suggest that administrative traditions can be linked to how power and complexity are established in collaboration, and a strong collaborative history may be linked to more acceptance of power imbalance. Due to the ambiguity in defining risks, P6 focusing on collaboration history and risks did not have sufficient findings to make any conclusions.

In all, our findings show certain behavioural clustering related to differences in administrative traditions such as pragmatism, legal culture, level of formality and hierarchy which did play a pivotal role in the perception of project challenges. This was most visible with the Continental and Scandinavian traditions, which represented contrasting findings with challenges from political aspects more relevant for the former and technical aspects for the latter. However, similarities did emerge in the context of actually managing the projects, which weakens the applicability of administrative traditions for evaluating challenges. Collaborative history between actors within a digitalisation project appeared to be the strongest regarding the threat posed by power imbalances. It did not appear to be as consistent concerning risk or complexity. Future research should go more in-depth into the specific administrative traditions. With a limited number of cases, the possible inferences remain limited. Furthermore, throughout the study, the interviewees reflected on the mitigating role of strong leadership alongside pre-existing relations. While this study focused on the latter, the former went beyond the scope of the current study and is a key research topic for the future. It also enables us to learn more about the applicability of administrative traditions. To conclude, as digital solutions become the default approach to government modernisation and change, government collaboration in this field will also change and adapt to balancing both the need for collective and interoperable digital solutions, while also taking into consideration the unique needs and logic of individual actors and environment during the process.

NOTE

1. The TROPICO project received funding from the European Union's Horizon 2020 research and innovation programme under grant agreement No.726840. For more information: https://cordis.europa.eu/project/id/726840.

REFERENCES

Alford, J., and O'Flynn, J. (2012). *Rethinking Public Service Delivery: Managing with External Providers.* Palgrave Macmillan.

Ansell, C., and Gash, A. (2008). Collaborative governance in theory and practice. *Journal of Public Administration Research and Theory, 18*(4), 543–71.

Bach, T., van Thiel, S., Hammerschmid, G., and Steiner, R. (2017). Administrative tradition and management reforms: A comparison of agency chief executive accountability in four Continental Rechtsstaat countries. *Public Management Review, 19*(6), 765–84.

Barrett, M., Heracleous, L., and Walsham, G. (2013). A rhetorical approach to it diffusion: Reconceptualizing the ideology-framing relationship in computerization movements. *MIS Quarterly: Management Information Systems, 37*(1), 201–20.

Benz, A., and Zimmerman, C. (2011). Germany: Varieties of democracy in a federal system. In J. Loughlin, F. Hendriks, and A. Lidstrom (Eds.), *The Oxford Handbook of Local and Regional Democracy in Europe* (pp. 146–72). Oxford University Press.

Breaugh, J., and Hammerschmid, G. (2020). Understanding public servants from a global perspective. In H. Sullivan, H. Dickinson, and H. Henderson (Eds.), *The Palgrave Handbook of the Public Servant* (pp. 1–20). Springer International Publishing.

Breaugh, J., Rackwitz, M., and Hammerschmid, G. (2023). Leadership and institutional design in collaborative government digitalisation: Evidence from Belgium, Denmark, Estonia, Germany, and the UK. *Government Information Quarterly, 40*(2), 101788.

Bryson, J., Crosby, B., and Stone, M. (2006). The design and implementation of cross-sector collaborations: Propositions from the literature. *Public Administration Review, 66*, 44–55.

Chen, C.A., and Bozeman, B. (2012). Organizational risk aversion: Comparing the public and non-profit sectors. *Public Management Review, 14*(3), 377–402.

Chen, Y., and Lee, J. (2018). Collaborative data networks for public service: Governance, management, and performance. *Public Management Review, 20*(5), 672–90.

Chen, Y., Hu, L., Tseng, K., Juang, W., and Chang, C. (2019). Cross-boundary e-government systems: Determinants of performance. *Government Information Quarterly, 36*(3), 449–59.

Choi, T., and Robertson, P.J. (2014). Deliberation and decision in collaborative governance: A simulation of approaches to mitigate power imbalance. *Journal of Public Administration Research and Theory*, *24*(2), 495–518.
Crosby, B., & Bryson, J. (2010). Integrative leadership and the creation and maintenance of cross-sector collaborations. *Leadership Quarterly*, *21*(2), 211–30. https://doi.org/10.1016/j.leaqua.2010.01.003.
Costumato, L. (2021). Collaboration among public organizations: A systematic literature review on determinants of interinstitutional performance. *International Journal of Public Sector Management*, *2005*.
Damanpour, F., and Schneider, M. (2006). Phases of the adoption of innovation in organizations: Effects of environment, organization and top managers. *British Journal of Management*, *17*(3), 215–36.
Dawes, S., and Pardo, T. (2002). Building collaborative digital government systems. In W. McIver and A. Elmagarmid (Eds.), *Advances in Digital Government* (Vol. 26, pp. 259–73). Springer.
de Vries, H., Bekkers, V., and Tummers, L. (2016). Innovation in the public sector: A systematic review and future research agenda. *Public Administration*, *94*(1), 146–66. https://doi.org/10.1111/padm.12209.
DeSanctis, G., and Poole, M.S. (1994). Capturing the complexity in advanced technology use: Adaptive structuration theory. *Organization Science*, *5*(2), 121–47. https://doi.org/10.1287/orsc.5.2.121.
Edelenbos, J., and Klijn, E.-H. (2007). Trust in complex decision-making networks. *Administration and Society*, *39*(1), 25–50.
Fedorowicz, J., Sawyer, S., Williams, C., Markus, M., Dias, M., Tyworth, M., Gantman, S., Jacobson, D., Tomasino, A., and Schrier, R. (2014). Design observations for interagency collaboration. *Government Information Quarterly*, *31*(2), 302–16.
Flemig, S., Osborne, S., and Kinder, T. (2016). Risky business – reconceptualizing risk and innovation in public services. *Public Money and Management*, *36*(6), 425–32.
Gil-García, J. (2012). Towards a smart State? Inter-agency collaboration, information integration, and beyond. *Information Polity: The International Journal of Government and Democracy in the Information Age*, *17*(3–4), 269–80.
Gil-García, J., and Sayogo, D. (2016). Government inter-organizational information sharing initiatives: Understanding the main determinants of success. *Government Information Quarterly*, *33*(3), 572–82.
Granovetter, M. (1973). The strength of weak ties. *American Journal of Sociology*, *78*(6), 1360–80.
Granovetter, M. (1985). Economic action and social structure: The problem of embeddedness. *American Journal of Sociology*, *91*(3), 481–510.
Guy Peters, B. (2021). *Administrative Traditions: Understanding the Roots of Contemporary Administrative Behavior*. Oxford University Press.
Hartley, J., Sørensen, E., and Torfing, J. (2013). Collaborative innovation: A viable alternative to market competition and organizational entrepreneurship. *Public Administration Review*, *73*(6), 821–30.
Hinterleitner, M., and Sager, F. (2015). Avoiding blame – a comprehensive framework and the Australian home insulation program fiasco. *Policy Studies Journal*, *43*(1), 139–61.
Huxley, K., Andrews, R., Downe, J., and Guarneros-Meza, V. (2016). Administrative traditions and citizen participation in public policy: A comparative study of France, Germany, the UK and Norway. *Policy and Politics*, *44*(3), 383–402.

Joffe, H. (2003). Risk: From perception to social representation. *British Journal of Social Psychology*, *42*, 55–73.
Jones, S., and Hooper, T. (2017). New Zealand's ICT strategy: The Respective roles of senior and middle management in promoting collaboration and innovation. *Asia and the Pacific Policy Studies*, *4*(3), 484–95.
Kapoor, K., Ziaee Bigdeli, A., Dwivedi, Y.K., Schroeder, A., Beltagui, A., and Baines, T. (2021). A socio-technical view of platform ecosystems: Systematic review and research agenda. *Journal of Business Research*, *128* (July), 94–108.
Kattel, R., Lember, V., and Tõnurist, P. (2020). Collaborative innovation and human-machine networks. *Public Management Review*, *22*(11), 1652–73.
Klijn, E., and Koppenjan, J. (2016). *Governance Networks in the Public Sector*. Routledge.
Kuipers, B., Higgs, M., Kickert, W., Tummers, L., Grandia, J., and van der Voet, J. (2014). The management of change in public organizations: A literature review. *Public Administration*, *92*(1), 1–20.
Lember, V., Kattel, R., and Tonurist, P. (2018). Technological capacity in the public sector: The case of Estonia. *International Review of Administrative Sciences*, *84*(2), 214–30.
Lewis, J., Ricard, L., and Klijn, E. (2018). How innovation drivers, networking and leadership shape public sector innovation capacity. *International Review of Administrative Sciences*, *84*(2), 288–307.
Luna-Reyes, L., Gil-García, J., and Cruz, C. (2007). e-Mexico: Collaborative structures in Mexican public administration. *International Journal of Cases on Electronic Commerce*, *3*(2), 54–70.
McGuire, M., and Agranoff, R. (2011). The limitations of public management networks. *Public Administration*, *89*(2), 265–84.
Mergel, I. (2016). Agile innovation management in government: A research agenda. *Government Information Quarterly*, *33*(3), 516–23.
Meyer, R., and Hammerschmid, G. (2010). The degree of decentralization and individual decision making in central government human resource management: A European comparative perspective. *Public Administration*, *88*(2), 455–78.
Nasi, G., Cucciniello, M., Mele, V., Valotti, G., Bazurli, R., de Vries, H., Bekkers, V., Tummers, L., Gascó, M., Ysa, T., Fernández, C., Albareda, A., Matei, A., Savulescu, C., Antonie, C., Balaceanu, E., Juraj, N., Svidroňová, M., and Mikusova Merickova, E. (2015). *Determinants and Barriers of Adoption, Diffusion and Upscaling of ICT-Driven Social Innovation in the Public Sector: A Comparative Study across 6 EU Countries.* European Union report. https://www.researchgate.net/publication/280881010_Determinants_and_Barriers_of_Adoption_Diffusion_and_Upscaling_of_ICT-driven_Social_Innovation_in_the_Public_Sector_A_Comparative_Study_Across_6_EU_Countries/link/55ca15e808aea2d9bdcbf68e/download, accessed 25 September 2023.
Neumann, O., Matt, C., Hitz-Gamper, B., Schmidthuber, L., and Stürmer, M. (2019). Joining forces for public value creation? Exploring collaborative innovation in smart city initiatives. *Government Information Quarterly*, *36*(4), 101411.
Nooteboom, B., Berger, H., and Noorderhaven, N. (1997). Effects of trust and governance on relational risk author. *The Academy of Management Journal*, *40*(2), 308–38.
Orlikowski, W. (2007). Using technology and constituting structures: A practice lens for studying technology in organizations. *Resources, Co-evolution and Artifacts*, 255–305.

Orlikowski, W., and Gash. D. (1994). Technological frames: Making sense of information technology in organisations. *ACM Transactions on Information Systems*, *12*(2), 174–207.

Osborne, S. and Brown, L. (2011). Innovation, Public policy and public services delivery in the UK: The word that would be king? *Public Administration*, *89*(4) 1335–50. https://doi.org/10.1111/j.1467-9299.2011.01932.x.

Painter, M., and Guy Peters, B. (2010a). Administrative traditions in comparative perspective: Families, groups and hybrids. In M. Painter and B. Guy Peters (Eds.), *Tradition and Public Administration* (pp. 19–30). Palgrave Macmillan.

Painter, M., and Guy Peters, B. (2010b). The analysis of administrative traditions. In M. Peters and B. Guy Peters (Eds.), *Administrative Traditions in Comparative Perspective: Families, Groups and Hybrids* (pp. 3–13). Palgrave Macmillan.

Pardo, T. (2010). Digital government implementation: A comparative study in USA and Russia. *16th Americas Conference on Information Systems*, 1–10. http://aisel.aisnet.org/amcis2010/330.

Pardo, T., Gil-Garcia, J., and Luna-Reyes, L. (2011). Collaborative governance and cross-boundary information sharing: Envisioning a networked and IT-enabled public administration. In *The Future of Public Administration around the World: The Minnowbrook Perspective* (pp. 129–39). Georgetown University Press. https://press.georgetown.edu/Book/The-Future-of-Public-Administration-around-the-World, accessed 25 September 2023.

Picazo-Vela, S., Gutiérrez-Martínez, I., Duhamel, F., Luna, D., and Luna-Reyes, L. (2018). Value of inter-organizational collaboration in digital government projects. *Public Management Review*, *20*(5), 691–708.

Pollitt, C., and Boukaert, G. (2011). *Public Management Reform. A Comparative Analysis – New Public Management, Governance, and the Neo-Weberian State* (third edn). Oxford University Press.

Purdy, J. (2012). A framework for assessing power in collaborative governance processes. *Public Administration Review*, *72*(3), 409–17.

Quick, K., and Feldman, M. (2014). Boundaries as junctures: Collaborative boundary work for building efficient resilience. *Journal of Public Administration Research and Theory*, *24*(3), 673–95.

Raadschelders, J., and Whetsell, T. (2018). Conceptualizing the landscape of decision making for complex problem solving. *International Journal of Public Administration*, *41*(14), 1132–44.

Rackwitz, M., Hammerschmid, G., Breaugh, J., and Palaric, E. (2020). *Government Collaboration and Digitalisation: Comparative Case Studies on Collaboration Managment for Government Digitalisation and Public Sector Innovation.* https://tropico-project.eu/download/d6-3-government-collaboration-and-digitalisation-comparative-case-studies-on-collaborative-management-for-government-digitalisation-and-public-sector-innovation/, accessed 9 September 2023.

Rackwitz, M., Hustedt, T., and Hammerschmid, G. (2021). Digital transformation: From hierarchy to network-based collaboration? The case of the German 'Online Access Act.' *Der Moderne Staat – Zeitschrift Für Public Policy, Recht Und Management*, *14*(1–2021), 101–20.

Ring, P., and de Ven, A. (1994). Developmental Processes of cooperative interorganizational relationships. *The Academy of Management Review*, *19*(1), 90–118.

Slovic, P. (1999). Trust, emotion, sex, politics, and science: Surveying the risk assessment battlefield. *Risk Analysis*, *19*(4), 689–701. doi: 10.1023/a:1007041821623.

Timeus, K. (2018). Passing the buck? How risk behaviours shape collaborative innovation. In T. Bach and K. Wegrich (Eds.), *The Blind Spots of Public Bureaucracies and the Politics of Non-Coordination* (pp. 151–70). bPalgrave Macmillan.

Timeus, K., and Breaugh, J. (2020). What's in it for me? How blame and credit expectations affect support for innovation. *Public Policy and Administration, 38*(2), 133–264.

Torfing, J. (2019). Collaborative innovation in the public sector: The argument. *Public Management Review, 21*(1), 1–11. https://doi.org/10.1080/14719037.2018.1430248.

Torugsa, N., amd Arundel, A. (2017). Rethinking the effect of risk aversion on the benefits of service innovations in public administration agencies. *Research Policy, 46*(5), 900–10.

Verhoest, K., Verschuere, B., and Bouckaert, G. (2007). Pressure, legitimacy, and innovative behavior by public organizations. *Governance, 20*(3), 469–97.

Wegrich, K. (2019). The blind spots of collaborative innovation. *Public Management Review, 21*(1), 12–20.

6. The coordination of smart cities: insights from a cross-case analysis on the implementation of smart city strategies

Maike Rackwitz and Dries van Doninck

INTRODUCTION

Cities around the world strive to strengthen digital innovation, attract economic activity, and create public value by becoming 'smarter' (Estevez et al., 2021; Meijer and Bolívar, 2016; Neumann et al., 2019). As intersections of economic, cultural, and social activity, they have been at the forefront of exploring the potential of technology to tackle increasingly complex societal problems (Barber, 2013; Drapalova and Wegrich, 2020; Landry, 2006).

While initial approaches to capturing this movement – such as Dutton et al. (1987) recognise only one networked city (Wired City) – a smart city today is regarded more as a gradual concept where cities are not 'smart or dumb', but their *smartness* is determined by 'its capacity to attract and mobilise human capital in collaborations between various actors through the use of information and communication technologies' (Meijer and Bolívar, 2016, p. 398). Put differently, what today constitutes 'smart' is, in addition to the pure networking of information and communication technology (ICT), the pluralism of actors and technologies like artificial intelligence, self-learning algorithms, and the internet of things. This makes control and thus the public management of smart city processes a particularly complex challenge (Ruhlandt, 2018). In this regard, it is often emphasised that competencies on the part of the city administration need to be strengthened to adequately face these challenges, and governance must become smart as well *(smart city governance)*. Only this way would local actors be able to comply with tasks and make independent decisions for the good of their city (Bolívar, 2016). As a result of the plurality of actors and high-end technologies, smart cities could be regarded as prime examples to study collaboration in a digital transformation context.

Although there seems to be a normative consensus on the mutually positive effects of collaboration as 'ends and 'means' of smart city initiatives, empirically based studies on network dynamics from an intra- and intergovernmental perspective remain scant, with an even greater need for systematic cross-national research (for some exceptions see, e.g., Pereira et al., 2017; Sancino and Hudson, 2020).

While the collaboration rhetoric in the digital realm has gained scholarly attention in recent years (Anthopoulos and Reddick, 2016; Mergel, 2016; Neumann et al., 2019), it often neglects the fact that stakeholder interaction can take on multiple facets, with collaboration being only one of them. From this perspective, it is questionable whether the frequent use of the analytical concept 'collaboration' adequately captures what is meant by navigating collective action towards desired smart city outcomes. However, because different intensities of partnerships may have distinct demands required for their governance, a more nuanced understanding is important for finding targeted public management interventions to guide smart city dynamics effectively. We take up this ambiguity in the literature and seek to explore the question: How does collaboration manifest in smart city networks?

To address this question, we draw on the theoretical notions of the 3C concept (e.g., Keast et al., 2007), which recognises varying degrees of partnerships along, as we argue, three key dimensions: interdependence, trust, and commitment. This helps us find evidence of cooperation, coordination, or collaboration in joint smart city efforts, ranging from the least to the most integrated and connected form. To this end, we conduct a comparative case study using archival data and semi-structured interviews. For the empirical setting, we select cases based on their similarities: the Belgian city of Antwerp and the German city of Darmstadt, both embedded in a distinct *Rechtsstaat* tradition and nationally known as digital pioneers, to investigate how they strategically proceed to master (novel) tasks of joint public management. This is done to hold the institutional logics constant to some extent, to isolate the role of project governance and its arrangements in dictating partnership dynamics (Wang and Feeney, 2016), while also exploring the extent of smart city governance variations across Europe.

This chapter is organised as follows. First, the theoretical framework is outlined; we elaborate on three intensity degrees of partnerships using the 3C framework and its core dimensions. Second, we move to our empirical analysis as we present the genesis and partnership arrangements of our two case studies, to which the framework is subsequently applied. Finally, we provide a comparative discussion of the results.

THEORETICAL FRAMEWORK

The '3C' Framework and Intergovernmental Relation

'Collaboration' is a term commonly applied in the context of smart cities to describe a set of organisational actors working together to achieve joint goals (e.g., Grossi et al., 2020; Meijer and Bolívar, 2016; Neumann et al., 2019; Pareira et al., 2017). This 'smart collaboration' is, in theory, often attributed to high-intensity interaction between actors, entailing a pronounced transformation (Meijer and Bolívar, 2016) through 'collective learning' (Meijer and Thaens, 2018, p. 367) and 'promoting communication, interaction, ... participation in decision making and direct democracy' (Pereira et al., 2018, p. 144). In contrast, Dameri and Benevolo (2016) and Gil-Garcia et al. (2015) highlight the various types of partnerships at play in smart city environments, such as networks, engagements or collaborations, which involve the activities of sharing, communication, and integration and, thus, surpass mere collaborative components. The sparse empiricism on smart collaboration echoes this discrepancy. In their study on UK smart city initiatives, Sancino and Hudson (2020), for instance, aim to investigate 'collaborative projects' (p. 706). However, their case selection reflects loose arrangements that are intended to facilitate platform interaction yet that have significant communication barriers. Moreover, in most smart city studies, 'collaboration' refers to relationships between governmental and non-governmental organisations (e.g., Meijer and Thaens, 2018; Wang et al., 2018; Zvolska et al., 2019), rarely examining whether this stands up to the test of 'true' collaboration, and even less so, how this plays out at other structural levels of smart city efforts within and between governments.

In the collaboration literature, strictly speaking, 'collaboration' refers to a specific form of working together (Huxham, 2000) and does not consider the varying degrees of actor commitment, engagement, and integration. In practice, however, partner-like relationships can be broken down into different facets of how intensively people work together and what they endeavour to do (Keast et al., 2007). Reflecting this, the '3C' concept has emerged, referring to 'cooperation, coordination, and collaboration' with its implications extending well beyond semantics. In the literature, there is a growing tendency to use 'cooperation', 'coordination', and 'collaboration' interchangeably or subsume them under each other. In fact, they can be placed on a horizontal continuum of increasingly intense partner interactions within a network (Costumato, 2021; McNamara, 2012).

Accordingly, each step along the continuum involves the stronger integration of 'mission and tasks, risks and rewards, and authority and accountability'

(Mattessich et al., 2001, p. 61; Thomson and Perry, 2006) and consequently greater interdependence between partnering organisations (Elston et al., 2018). We will position each of the Cs on this continuum, which allows us to work out and delineate them according to their core dimensions, and ultimately make the case for a more nuanced understanding of partnerships within smart cities.

Cooperation

Cooperation can be understood as a starting point for further, more intense interaction usually accompanied by low structural linkage and, thus, high fragmentation. Information sharing is the preferred choice of exchange and formal, binding agreements are dispensed with (McAllister and Taylor, 2015). This is often associated with the actors' low commitment and low willingness to relinquish their autonomy, which can partly be explained by the anticipated risk that others will behave opportunistically (Keast et al., 2007). Therefore, a lack of or underdeveloped trust plays a decisive role in this form of partnership (O'Leary and Bingham, 2007). However, the risk and rewards of interacting are comparatively modest (Sedgwick, 2016).

Coordination

Coordination becomes necessary when specialisation and more formalised, repeated interaction is desired (De Pourcq and Verleye, 2021). It has a more instrumental function, presupposing a basic concern between partners to act in concert by structurally adapting to each other (Ansell, 2000). While this may involve the pursuit of a common predetermined goal, it is task-oriented and does not necessarily require cultural adjustment or a loss of individual autonomy of the partnering organisations. The partners exchange not only information but also resources, which entails greater risk and potentially greater benefit from the partnership (Keast et al., 2007; Mattessich et al., 2001).

Collaboration

Collaboration refers to the most intensive and comprehensive mode of interaction that promotes mutual dependence, goal alignment, and joint planning and action, often accompanied by a structural and cultural blurring of organisational boundaries (Gray, 1989; Sedgwick, 2016; Thomson et al., 2009). Partners are more proactive and 'whole-hearted' (Keast et al., 2007, p. 17), involving a higher level of trust. Simultaneously, this includes a more decentralised form of power-sharing where all members are responsible for controlling the network. With it, the risk and the potential reward magnify (Ansell and Gash, 2018; Whelan, 2015).

The 3C Dimensions: Interdependence, Trust, and Commitment

A closer inspection of 3C concepts reveals that they differ analytically along three interrelated dimensions: interdependence, trust, and commitment (see Figure 6.1 for the horizontal partnership intensity continuum). We will briefly outline what the collaboration literature understands through these dimensions and finally use the framework to compare our two cases.

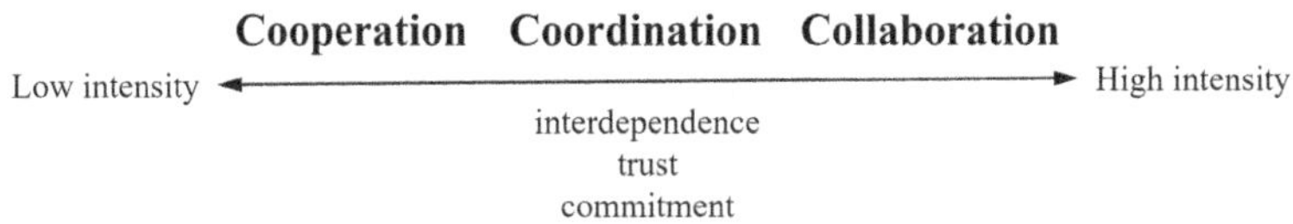

Figure 6.1 Horizontal partnership intensity continuum

Interdependence is understood as the extent of connections between elements of a system or 'nodes connected to other nodes' within partnerships (O'Leary and Bingham, 2009; Whelan, 2015). Hence, interdependence implies connectedness and decreasing autonomy of partners as it relates to the position that actors take within the structural linkages or ties that connect them (Lewis et al., 2016). It also refers to the types of assets shared. These can be tangible or intangible, ranging from the sharing of information to resources to power. Interdependence occurs whenever 'one actor does not entirely control all of the conditions necessary for the achievement of an action, or for obtaining the outcomes desired from the action' (Pfeffer and Salancik, 1978, p. 40).

Similarly, building *trust* stimulates partnerships by reducing the transaction costs of oversight and more formalised governance (Ring and van de Ven, 1994) and, thus, the risk of adverse strategic behaviour (Kwon and Suh, 2004). Trust is broadly defined as 'expectations that any information or other resources provided will not be used by the other in ways that could do harm to oneself' (Wiedner and Ansari, 2019, p. 200). This effect is potentially enhanced by face-to-face dialogue and frequent exchanges (Ansell and Gash, 2008; Lasker and Weiss, 2003).

Trust has been shown to be positively related to *commitment* (Gray, 1989), the willingness of partners to invest in a relationship. More precisely, inter-party commitment, or attachment, can be defined as an 'inertial or binding force between exchange partners that can lead to the maintenance of an existing relationship to the exclusion of alternatives' (Seabright et al., 1992, p. 126). A strongly committed partnership includes partners who recognise mutual dependence and exert mutual control while empowering each other to initiate actions (Ansell and Gash, 2008). Like trust, commitment is attributed to vulnerability as it is highly contingent upon partners' agreement on values and goals (Talay and Akdeniz, 2014).

However, this continuum, with its dimensions outlined, does not imply a normative view that assumes that greater intensity is accompanied by an increase in partnership effectiveness. In fact, for instance, while interdependence can be seen as the key rationale for forming partnerships (Gray, 1989), too much interdependence can drive up transaction costs and increase the perceived or objective management burden resulting from the partnership (Elston et al., forthcoming). Overdependence can also create tensions as organisations tend to value their autonomy (Seabright et al., 1992). Additionally, while trust is a desirable prerequisite for partnerships (Huxham and Vangen, 2005), overreliance in relationships may lead to negligent behaviour. This risks exploitation and brings strategic disadvantages, especially in opportunistic environments. In a similar vein, over-commitment in the face of disconfirming evidence or changes in resource fit can lead to holding on to something shown to be ineffective (Jarvenpaa and Majchrzak, 2016; Seabright et al., 1992).

As commitment and trust are built through interaction and joint experiences of success (Ansell and Gash, 2008; Vangen and Huxham, 2003), collaborative intensity can be assumed to vary within partnerships. By engaging in joint efforts over time, actors can get to know each other and build up expectations of reciprocity and shared meaning, lifting the intensity level of the partnership (Liu and Zheng, 2018). Following Ansell and Gash (2008) and Emerson et al. (2012), interactions within a partnership are also influenced by structural conditions including the establishment of hierarchies, process rules or arenas for interaction. As a consequence, we expect the intensity of interactions to vary by temporal and structural features within partnerships.

By applying this relevant distinction of working in partnerships to our empirical context of two Western European smart city networks, we explore along the 3Cs and their dimensions the extent to which they involve more intensive collaboration or represent alternative forms of partnership, such as coordination. This considered, we extend the initial research question, asking: What level of partnership intensity manifests in smart city networks and how do time and structural differences affect this level of intensity?

To this end, we drew on document analysis of items including case-related strategic documents, meeting agendas, minutes, and newspaper articles in both countries. Semi-structured interviews were then used to verify initial findings and gain complementary insights from those involved in setting up or maintaining the smart city cases. This resulted in a total of 12 interviews, six in each country, with bureaucrats (non-politicians) representing different interests and positions to ensure a balanced picture (see Table 6.1 for interviewee details). Interviews were each conducted by native interviewers at participants' workplaces between September 2019 and March 2020 and lasted up to 90 minutes.

Table 6.1 *Overview of interviewees and their professional positions with numbering in brackets referred to in the main text*

Case	Country	Senior mgmt.	Middle mgmt.	Total
Antwerp's smart city policy	Belgium	3 [A1:A3]	3 [A4:6]	6
Digitalstadt Darmstadt	Germany	4 [D1:D4]	2 [D5:D6]	6
				12

They were recorded and transcribed verbatim and coded along with the selected documents using MAXQDA software. This served to systematise the rich material in line with our research aim.

CASE INTRODUCTION

In the following sections, we give a brief description of the background and context of the cases under study. The cases selected are smart city projects in two medium-sized Western European cities: the smart city policy of Antwerp in Belgium and the Digitalstadt Darmstadt project in Darmstadt, Germany (see Table 6.2 for key case characteristics). Both cases can be characterised as Anthopoulos's (2017) model 1, where the city itself or one of its outsourced entities takes over the management of the smart city – a common model used in other cities such as Amsterdam, Barcelona, or Vienna.

Similarities in System Context

As mentioned, both cases represent leading digital cities within their countries. Hence, some similarities between the cases can be found in their system context. The first similarity is both cities' economic position within their region or country. Antwerp is the largest city in the Belgian Flemish region. In 2015, it won the first Global Startup Nations Award for Local Policy Leadership (TakeOff Antwerp, 2015). In traditional economic sectors like logistics and retail, Antwerp has one of the most important concentrations of economic activity due to its central location in Europe, close to numerous European highways and railroads. Additionally, the Port of Antwerp is one of the largest seaports in Europe and hosts the largest integrated chemical cluster in the world. Located close to the Rhine River and between big industrial cities like Frankfurt-am-Main and Mainz, Darmstadt is situated in the economically prosperous Hesse region. The Rhine-Main-region is dominated by activities in manufacturing, pharmacy, chemistry, and biotechnology. In 1997 Darmstadt was awarded the title of *City of Science*. Several scientific institutes are located in Darmstadt including the European Space Operations Command (ESOC) of the European Space Agency. Further, the Technical

University of Darmstadt (TU Darmstadt) is one of the most important research universities in the country. Darmstadt has also been awarded for its efforts in digital transformation. Additionally, the city has been honoured as the most sustainable German city four years in a row by a well-established magazine called *Wirtschaftswoche* (Bitkom e.V., 2019).

A second similarity in the cases' system context can be found in the political environment of both cities. Both Germany and Belgium are federal states with smart city competencies regarding innovation and digital transformation mainly located at the regional and local levels. Additionally, both municipalities could rely on strong support from the regional level. In 2012, a centre-right coalition took office in Antwerp with an economically liberal agenda focussing on innovation and economic development. In 2014, a coalition represented by the same political parties took office in the Flemish government with a similar economic agenda. The Flemish region invested in digital transformation in its coalition agreement and put this into practice by launching the Smart Flanders programme, in partnership with IMEC, around the same time the city of Antwerp launched its local Smart City Programme in partnership with IMEC. Darmstadt is a so-called *kreisfreie Stadt*, which means that it directly falls under the regional *Länder* policy level, which is similar to Antwerp's position within the multi-level structure of the Belgian state. Darmstadt found support at the regional policy level for its smart city projects as the Hesse region declared Darmstadt a model municipality (*Hessische Modellkommune*), which subsequently became the support of €5 million for the smart city projects.

Antwerp's Smart City Policy

The smart city policy of Antwerp was formally kicked off in 2016 when the city council signed a covenant with IMEC, a research centre renowned for its work in digital transformation and smart cities. The general strategy for smart city projects in the city revolves around five 'building blocks': a digital transformation of the city administration's front and back office, the creation of a single digital platform, the creation of Europe's biggest open laboratory, an open data policy, and an ecosystem for digital innovations. The most important output of the smart city policy leading up to the time of data gathering at the end of 2020 has been several digital innovation projects conducted in the open laboratory also referred to as the 'smart zone'.

IMEC and the city of Antwerp have a quite elaborate history of joint working, as they have been co-partners in several research projects funded through the European Commission's Horizon 2020 programme or innovation programmes of the Flemish regional government. A third important actor in the smart city policy is Digipolis. Digipolis, at the time of the inquiry, was an 'inter-municipal association', which is an autonomised legal structure to which governments can delegate certain public services. Digipolis was responsible for all IT-related activities of the city of Antwerp, including digital infrastructure.

Regarding the governance arrangements of the partnership, the backbone of the smart city activities is a covenant between the city of Antwerp and IMEC in which they agree on a central strategy for the smart city policy of Antwerp and on the covenant budget, for which they each contribute equally (650,000 euros for the duration of the covenant). After an evaluation of the first years of the covenant, an addendum was added in which a formal governance structure of smart city projects was implemented. A schematic representation of this governance structure can be found in Figure 6.2.

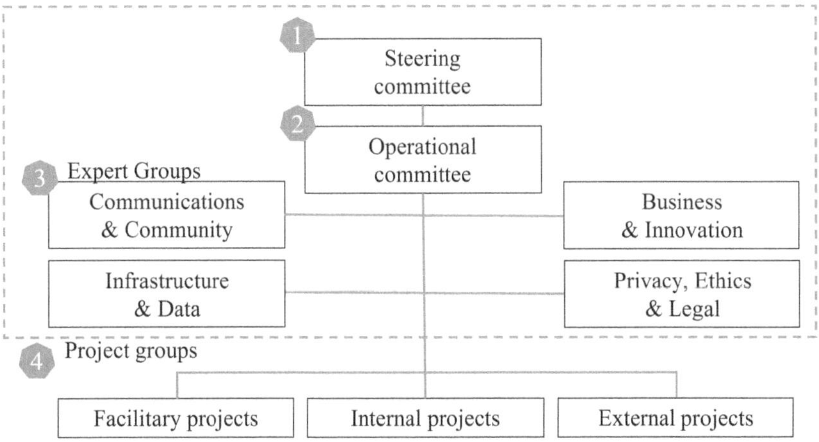

Source: Stad Antwerpen (2020).

Figure 6.2 *Governance structure of Antwerp's smart city projects and general policy*

The governance structure is made up of hierarchical levels where different roles of governing different parts of smart city activities were assigned to different partners within the governance structure. For example, at the steering committee level, the global strategy of the smart city is laid down and decisions about the covenant budget are made. Members of the city administration, the mayor, and the alderman of Economy and IMEC are represented. On the second level is the 'operations team'. It is an in-between level where the decisions of the steering committee are translated into concrete actions for the projects. Here, the members of the steering committee are represented, along with operational managers of both IMEC and the city administration. Often, representatives of ongoing projects are invited as well. Important to note is that this body has no formal decision-making power, but they do prepare the agenda for the steering committee. Third, expert groups debate on specific, thematic issues that arise in the projects. The fourth level comprises the actual project teams, where the different smart city projects are executed and implemented.

Digitalstadt Darmstadt

Our second case, the Digitalstadt Darmstadt smart city project developed from 2016 onwards. After winning the 'Digitale Stadt' competition in 2017, several stakeholders engaged in partnerships with the city of Darmstadt. A total of 32 funded projects were set up in accordance with an overarching smart city strategy. This strategy was already in place before the Digitale Stadt competition but as the competition approached, it was aligned with the city's existing economic strategy (HEAG, 2018). The projects encompass 14 areas of digital innovation: administration, mobility, trade and tourism, health, culture and Industry 4.0. Examples of projects are intelligent traffic light control, smart parking, a House of Digital Media Education, multimodal mobility and patient data networking. These projects are developed in partnership with several external partners like the University of Darmstadt, Fraunhofer Research Institutes, companies, or the city administration.

Like in Antwerp, the activities of Digitalstadt Darmstadt are also structured in a separate governance structure. In 2018, a limited liability company was founded under private law. This 'GmbH' was established as a subsidiary outside the core administration's structure. Digitalstadt Darmstadt GmbH is a 100 per cent subsidiary of the city of Darmstadt, there is a 95 per cent funding quota from external donors (5 per cent own funds). Bitkom with its private sponsors as well as the Hessian Ministry, therefore, play a crucial role in how the projects evolve. Besides these, the broader network consists of 50–70 'strong partners' (Digitalstadt Darmstadt, 2018), from politics, business, and science such as the Technical University of Darmstadt or several Fraunhofer research institutes, who have all contributed additional financial or technical support for individual project implementations.

Quite similar to the Antwerp case, the governance structure is made up of different hierarchical levels where different roles regarding governance are separated across actors. The strategic committee comprises top-level decision-makers with no direct connection to the projects on the ground. These are the mayor, the funding office of the Hessian Ministry for Digital Affairs, the Chief Digital Officer of the City of Darmstadt, the two managing directors of the GmbH, as well as the head of the Department of Urban Development and Economics and the CEO of HEAG, who were already actively involved in the competition participation. The second level is comprised of two exec-

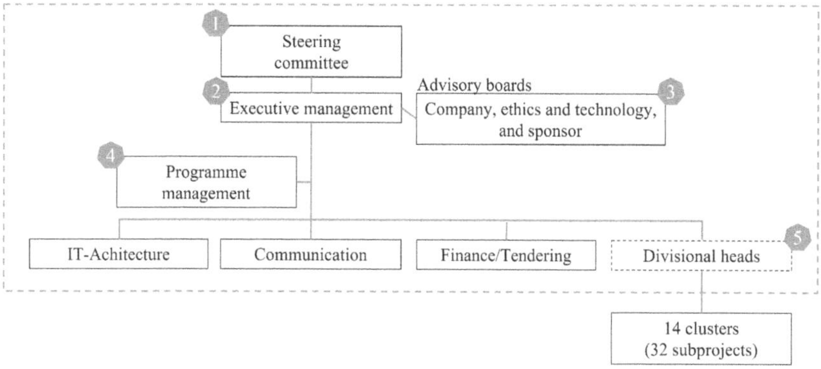

Source: Authors' illustration based on Roland Berger (2018).

Figure 6.3 Governance structure of Digitalstadt Darmstadt

utive managers employed by the GmbH to provide the expert groups with information and serve as a link between them and the steering committee. Additionally, they are responsible for leading the internal organisation of the GmbH's own departments such as programme management, IT, communication, or tenders (represented by level 4 in Figure 6.3). Lastly, the divisional heads make up another hierarchical level and they represent the project groups towards the executive management and the GmbH's departments.

ANALYSIS AND RESULTS

In this chapter, we look at two cases of smart city partnerships to explore the different guises of interactions in digital transformation networks. Smart city initiatives in two cities nationally leading in digital innovation were analysed through a document study and qualitative analysis of 12 interviews (six for each case) with non-political practitioners active in different roles and positions within the partnership.

Our analysis reveals that stakeholder interaction within smart city innovation partnerships comes in many shapes and is more nuanced and dynamic than often assumed in current scholarship on collaboration in digital transformation. We find two main dynamics in our data along which the level of interdependence, trust, and commitment varies and, as a consequence, the intensity of interactions varies.

Table 6.2 Key characteristics of the smart city cases

Case	Smart City Policy, Antwerp	Digitalstadt Darmstadt, Darmstadt
Country	Belgium	Germany
State structure	Federal	Federal
Project duration	2016–2019/no ongoing due to renegotiation	2018–21/now ongoing due to new funding
City size (population as of 2019)/ structure	525,936/monistic	159,103/dualistic
Primary motives	Efficiency and quality improvement/voluntary	Quality improvement/voluntary
Main partner responsible	City countil and IMEC	Digitalstadt GmbH

Temporal Dynamic

First, we distinguish a *temporal dynamic* where the intensity of partnership interactions varies across time.

The origin of both cases lies in interactions with a high level of intensity of a small, select group of influential collaborators. In Darmstadt, a close group of assertive and influential people teamed up to prepare Darmstadt's participation in the Bitkom competition. This 'inner circle' [D3] including the mayor, the chairman of the board of HEAG Holding AG, and the head of the urban development and economy department was driven by high levels of commitment and interdependence, the interactions between these collaborators had a high level of intensity that resembled collaboration (Keast et al., 2007). As they prepared a joint bid for the competition, mutual alignment of goals and joint action were necessary to outline implementation structures and to enthuse the local ecosystem. These horizontal interactions, with weekly meetings over an extended period of time, resulted in winning the competition and kickstarted the Digitalstadt Darmstadt project.

In Antwerp, the partnership originated in a rich prehistory of collaboration between a limited group of core partners. IMEC, the Department of Innovation and Digipolis had partnered in various digital innovation and smart city projects in the past, funded by a variety of government programmes including the European Commission's Horizon 2020 and the Smart Flanders programme. The interdependency that became apparent, and the trust developed in these projects, can be seen as a breeding ground that created an opportunity to engage in a more intense form of interaction that resembles elements of coordination, with a more formalised interaction through the signing of a joint covenant, and collaboration, with the alignment of goals and the planning of joint action that entailed some blurring of structural and eventually cultural organisational boundaries. 'We did many things from Antwerp, we did many things in Antwerp. We did some things with Antwerp already. So, why not do these smart city activities together with the city', one core partner recalled [A1].

In both cases, however, the intensity of the departure of the partnerships was only temporary. As the networks grew and external partners were included, loss of autonomy, conflicting interests, and increasing risk became more salient. Many projects in both cases, therefore, started off slowly, with very little output in the beginning. In Antwerp, time was needed for goal alignment and the development of mutual understanding, where trust and commitment were at a lower level resulting in less intense interactions resembling cooperation, mainly revolving around the exchange of information, and characterised by a reluctance to exchange other resources. 'The hard thing was that we had different finalities, I think. And this came through to the working floor as we asked how we will set up these projects?' 'I have to say that the core partners did find each other eventually, but in the beginning, it was hard', a core partner recalled [A2]. The interdependence created in the period of interactions with a high intensity revealed value conflicts among the more peripheral actors in the network.

In the project teams, the intensity of interactions was also impeded by a lack of commitment from the city departments. As they did not have the capacity, nor the knowledge to fully commit to the projects, goal alignment and joint action were impeded [A6]. The intensity level of interaction, therefore, resembled cooperation. In Darmstadt, the fast growth of the network also required the projects to slow down in the beginning. Constant dialogue and communication were required to sustain the Digitalstadt network, however, this also resulted in tensions with managing the projects. A step-by-step approach proved to be effective, although time-consuming, to manage the different complexities that come with a large and diverse network. In other words, while there were not necessarily issues of trust and commitment, connections between the different parts of the network had to grow and this limited the possibilities for joint action and goal alignment. In this phase, the interactions between the partners in the network were 'stuck' at the level of cooperation. 'First, think in large dimensions and then break it down into actionable measures', an interviewee explained [D2].

The above-mentioned *growing pains* could be resolved by continuing interactions between partners, but by lowering the intensity of the interactions. In Antwerp, for example, tension arose about the use of (public) data within smart city projects. While these are integral aspects of most smart city innovation projects, the use and reuse of public data are also highly controversial, especially for public organisations [A4]. Further, data standards are very important to enable the use of data for smart city innovations. These tensions became more salient as projects moved on and as the intensity of the partnership and the exchange within developed, which caused many projects to stall. By implementing a data charter and the aforementioned formal governance structure with expert groups who could advise on these issues, these issues were eventually resolved [A1; A2; A4]. Organisational boundaries were re-emphasised (e.g., as a data supplier and a data user), which created clarity and reduced the risk of committing to the projects. The first target agreements in the Digitalstadt Darmstadt projects were also respectful of organisational identities and cultures. Later on, regular round tables were introduced where core partners of the smart city partnership could meet with external players in the ecosystem to generate and develop ideas which caused 'the network to get stronger', as one respondent formulated [D4].

Structural Dynamic

Aside from interaction intensity varying across time, we also find a *structural dynamic* where the intensity of partnership interactions varies as a function of elements within the partnership's structure like roles or hierarchical levels.

For instance, the position of the interaction within the governance structure of the network was of influence for the intensity of the interaction. For example, core partners who are represented at the higher hierarchical level find it easier to develop more intense modes of partnership interaction. In Darmstadt, this is partly attributed to the visionary mayor's leadership style. He deliberately installed a culture of trial and error by creating spaces for mutual learning and allocating time for personal interaction. As one interviewee emphasised, unlike in the city administration, 'we call each other by our first names' [D5]. High levels of interdependence foster more intense interaction, which creates opportunities to grow trust and commitment. Similarly in Antwerp, the core partners also put forward openness and transparency towards each other: 'I think we tried to have the core partners represented in everybody and every project. That is one thing. Another is to be as open as possible about the lessons learned and to make them available for each partner' [A3]. Core partners are generally included in the governance structure's top hierarchical levels. They, therefore, deal with issues that are strategic, broad and more encompassing of the entire smart city. This allows their exchanges to be more visionary while engaging in more abstract thinking and requiring a strong sense of commitment.

In lower hierarchical levels, however, like in Antwerp's expert groups and project teams or Darmstadt's project teams or programme management groups, the discussions are more operational and *hands-on*. In this environment, conflicting interests and risks are more salient and concrete. An example from the Antwerp case was a test case with smart streetlights. While this certainly fits into the general goal of enabling smart technologies in the urban environment, there are several issues like altering the electric circuit of street lighting or the tramway overhead line to install a smart camera [A6]. In other words, when getting into the nitty gritty of such a project practical problems are created which have to be solved in a way compatible with the normal service delivery of the city administration [A5]. The commitment of city departments towards the smart city projects was low, as their focus is on the daily service delivery in the city. This prevents specialisation and joint action and limits the intensity of interactions within the project teams. Projects are then limited to experiments and do not evolve to a level of mutual goal development or the creation of a joint vision.

Another element creating variation in interaction intensity as a result of structural conditions is the capacity the individuals have to participate in smart city initiatives. This is generally higher when moving towards the upper hierarchical levels of the governance structure. In Darmstadt, for example, the GmbH Digitalstadt Darmstadt has employees on its payroll, including two programme managers. Their sole occupation is to manage the smart city programme of the city. In Antwerp, the city administration has a smart city director, as does IMEC. Both are represented in the steering committee with the task of developing the smart city policy in Antwerp. In the project teams, however, civil servants of city administrations or employees of private companies are included in the collaboration. They are indispensable to executing smart city innovation projects in a real-life *living* context, but they also have many other tasks within their job as a civil servant, for instance, in the Department of Mobility, or as an employee of a private company. This is also acknowledged by a respondent in the higher hierarchical level: 'It is not always clear for them what they have to win. They have to put a lot of time and effort into something that won't achieve their goal directly' [A2],

CONCLUSION

The purpose of this chapter was to shed light on how different intensity levels of collaborative interactions manifest in smart city innovation partnerships and how temporal and structural differences affect this intensity. In doing so, we sought to better understand whether these partnerships – as commonly assumed – involve more intensive collaboration between partners or instead represent alternative interactive forms such as coordination and what this implies for public management practice, drawing on the theoretical notions of the 3C concept (e.g., Keast et al., 2007). We focused on two cases in cities that are regarded as digital pioneers, the cities of Antwerp in Belgium and Darmstadt in Germany and conducted six interviews and a document study for each case. Both cases have relatively similar elements in their system context, which allowed our analysis to focus on the interactions between different partners within the network to explore the different ways in which these interactions can manifest in smart city innovation partnerships. In this concluding section of our chapter, we highlight our findings and outline implications for practitioners and researchers.

The analysis revealed that in the cases under study smart city networks are not clearly collaborative, as lower intensities of partnership were primarily observed. The answer to the question of how collaboration within smart city networks manifests is, therefore, more nuanced and indicates a varying intensity of interactions depending on the project phase and structural level. More so, our results reveal a temporal dynamic and a structural dynamic in the intensity of interactions between the members of the partnership (e.g., Ansell and Gash, 2008).

These findings have several implications for both practitioners and researchers of smart city projects. First, it is important to recognise the dynamic nature of partnership intensity. Collaboration is resource-intensive and creates several challenges that must be overcome. Therefore, interactions within the network cannot always be highly intense. Rather, periods of intense collaboration are needed to lift the network into a different phase. For example, episodes of intense collaboration can lead to the formulation of joint goals or a compelling vision which can kickstart the formation of a formal network. However, a new phase of partnership creates new challenges, and the engine of collaboration can become overheated as complexities, interdependencies, and conflicting interests become more salient (Elston et al., forthcoming; Huxham and Vangen, 2005; Seabright et al., 1992). In this case, managers of smart city projects need to allow the partnership to slow down and gain some distance. This period of low intensity can support problem definition and the development of solutions like institutional design measures. These new, formal design elements can then serve as *handles* to increase the intensity again and continue working towards joint goals. Our results suggest periods of high, medium, and low intensity need to be alternated to resolve conflict and move forward.

A second dynamic in our findings is a structural one. We find that the intensity of partnership interactions within smart city networks is also a function of structural elements. In our results, we mainly find evidence to support the observation that intensity varies by position within the governance structure of the interaction. Interactions more easily evolve towards an intensity level of collaboration at the top hierarchical levels of the network, where strategic decisions are made, and discussions are likely to be more abstract and related to the overall objectives and vision of the network. In project teams, however, operational decisions are made, and these decisions are more prone to value conflicts and practical obstructions. Therefore, our analysis finds the interaction intensity at the project level to be more limited and needs to be designed with greater caution. Managers and scholars of smart city networks, therefore, have to be aware that a network can be diverse, and that different forms of partnership and different challenges can arise in different areas of the network (Emerson et al., 2012). It is also important to align the strategic and operational perspectives on the problem. A two-way exchange between the top and the bottom is important to make sense of the bumps in the partnership road at the different levels.

Our nuanced perspective on partnerships with associated modes and intensities that has emerged thus suggests that a broad or fuzzy understanding of smart cities as collaborations falls short of adequately assessing what is needed to navigate network dynamics towards desired digital outcomes. Rather, it appears that public management practices must be attuned to varying degrees of intensity. In line with Keast et al. (2007), the suitability of each of the 3Cs is about 'getting the right mix' (p. 10) across structural levels within the partnership and over time (see also Ansell and Gash's 2008 'contingency approach'). This yields an important takeaway point of our chapter, that collaboration may represent the most intensive form of partnership but does not come without costs. It is difficult to achieve the intensity level of collaboration, and it can be resource-intensive, but it is even more difficult to sustain it (see also Keast et al., 2007). Therefore, managers should not aim for high-intensity interaction all the time and at every stage in the partnership, but rather find a balance between intense collaboration and less intense interactions where there is room for regrouping and reconsidering one's own values, resources, and objectives.

It is also interesting to note that in both Antwerp and Darmstadt, the project outcomes were moderate at the time of the interviews, three or four years after the formal founding of the partnership. Although the programmes and, in the case of Darmstadt, the funding period had already advanced, very few of the subprojects had entered the concrete implementation or scale-up phase. In addition to regulatory restrictions, especially with regard to legal uncertainties in the digital field, this is also due to the underestimated effort or time required to build and maintain fruitful networks. Given the volatile nature of interdependence, trust, and commitment, this may involve fragile phases of cooperation and collaboration and, as shown, more stable coordination in the longer term.

Again, important to note here is that collaboration, although the most intense form of interaction, is not always the adequate choice to achieve the desired partnership outcomes. Rather, intensive, shorter periods of collaboration with smaller groups of the right partners should be pursued only when most appropriate. During these short stints of intensive interactions, managers should build a trusting environment with regular yet informal exchange and strong commitment, while avoiding the exhaustion of partners' engagement and resources. In the long run, managers should also consider other, less intensive forms of interaction such as cooperation and coordination to ensure partnership resilience in the face of both intentional and unintentional fluctuations in staff, rationales, and resources.

In fact, the analysis is a snapshot from the beginning of 2020. The smart city projects and thus the interactions taking place may have intensified and weakened again in the meantime. In Darmstadt, for example, the planning of an overarching data platform was discussed, in the course of which additional joint structures, visions, and a resource pool are to be created. This is further supported by a newly generated grant of 13.3 million euros from the Federal Ministry of the Interior (BMI), which is to ensure the continuation of Digitalstadt's endeavours for the next seven years (BMI, 2020). In Antwerp, the learning outcomes were regarded as the most successful at the time of the interviews. A positive evaluation resulted in a renewal of the covenant, also implying a further investment of financial resources, in which expectations and goals for the partnership were delineated more sharply (Stad Antwerpen, 2020).

Building on this empirical work and following up on recent project developments would be a promising starting point for future analysis. This can help to better understand potential shifts in collaboration and derive practical implications for leadership and institutional design, considering partnership intensity as a key factor in finding appropriate mechanisms for effective smart city governance in Europe and beyond.

REFERENCES

Ansell, C. (2000). The networked polity: Regional development in Western Europe. *Governance*, *13*(2), 279–91.

Ansell, C., & Gash, A. (2008). Collaborative governance in theory and practice. *Journal of Public Administration Research and Theory*, *18*(4), 543–71.

Ansell, C., & Gash, A. (2018). Collaborative platforms as a governance strategy. *Journal of Public Administration Research and Theory*, *28*(1), 16–32.

Anthopoulos, L.G. (2017). *Understanding smart cities: A tool for smart government or an industrial trick?* (Vol. 22). Cham: Springer International Publishing.

Anthopoulos, L.G., & Reddick, C.G. (2016). Understanding electronic government research and smart city: A framework and empirical evidence. *Information Polity*, *21*(1), 99–117.

Barber, B.R. (2013). *If mayors ruled the world: Dysfunctional nations, rising cities*. New Haven, CT: Yale University Press.

Berger, R. (2018). *Strategie der Digitalstadt Darmstadt.* Darmstadt: Digitalstadt Darmstadt GmbH.

Bitkom e.V. (2019). *Smart-City-Atlas – Die kommunale digitale Transformation in Deutschland.* Berlin: Bitkom e.V.

BMI (2020). *Modellprojekte Smart Cities 2020. Gemeinwohl und Netzwerkstadt/ Stadtnetzwerk.* Berlin: Bundesministerium des Innern, für Bau und Heimat (BMI).

Bolívar, M.P.R. (2016). Mapping dimensions of governance in smart cities: Practitioners versus prior research. In *Proceedings of the 17th international digital government research conference on digital government research*, 312–24.

Costumato, L. (2021). Collaboration among public organizations: A systematic literature review on determinants of interinstitutional performance. *International Journal of Public Sector Management*, *34*(3), 247–73.

Dameri, R.P., & Benevolo, C. (2016). Governing smart cities: An empirical analysis. *Social Science Computer Review*, *34*(6), 693–707.

De Pourcq, K., & Verleye, K. (2021). Governance dynamics in inter-organizational networks: A meta-ethnographic study. *European Management Journal*, *40*(2), 273–82.

Digitalstadt Darmstadt (2018). *Ein Jahr nach dem Bitkom-Wettbewerb*. https://www.digitalstadt-darmstadt.de/news/ein-jahr-nach-dem-bitkom-wettbewerb/ (accessed 22 March 2023).

Drapalova, E., & Wegrich, K. (2020). Who governs 4.0? Varieties of smart cities. *Public Management Review*, *22*(5), 668–86.

Dutton, W.H., Blumler, J.G., & Kraemer, K.L. (1987). *Wired cities: Shaping future communication*. Macmillan.

Elston, T., MacCarthaigh, M., & Verhoest, K. (2018). Collaborative cost-cutting: Productive efficiency as an interdependency between public organizations. *Public Management Review*, *20*(12), 1815–35.

Elston, T., Rackwitz, M., & Bel, G. (forthcoming). Going separate ways: Ex-post interdependence and the dissolution of collaborative relations. *International Public Management Journal*.

Emerson, K., Nabatchi, T., & Balogh, S. (2012). An integrative framework for collaborative governance. *Journal of Public Administration Research and Theory*, *22*(1), 1–19.

Estevez, E., Pardo, T.A., & Scholl, H.J. (Eds.) (2021). *Smart cities and smart governance: Towards the 22nd century sustainable city* (Vol. 37). Springer Nature.

Gil-Garcia, J.R., Pardo, T.A., & Nam, T. (2015). What makes a city smart? Identifying core components and proposing an integrative and comprehensive conceptualization. *Information Polity*, *20*(1), 61–87.

Gray, B. (1989). *Collaborating: Finding common ground for multiparty problems*. San Francisco, CA: Jossey-Bass.

Grossi, G., Meijer, A., & Sargiacomo, M. (2020). A public management perspective on smart cities: 'Urban auditing' for management, governance and accountability. *Public Management Review*, *22*(5), 633–47.

HEAG (2018). *Durch Strategie zur Stadtwirtschaftskultur*. https://www.heag.de/durch-strategie-zur-stadtwirtschaftskultur/ (accessed 22 March 2023).

Huxham, C. (2000). The challenge of collaborative governance. *Public Management*, *2*(3), 337–57.

Huxham, C., & Vangen, S. (2005). *Managing to collaborate: The theory and practice of collaborative advantage*. London: Routledge.

Jarvenpaa, S.L., & Majchrzak, A. (2016). Interactive self-regulatory theory for sharing and protecting in interorganizational collaborations. *Academy of Management Review*, *41*(1), 9–27.

Keast, R., Brown, K., & Mandell, M. (2007). Getting the right mix: Unpacking integration meanings and strategies. *International Public Management Journal*, *10*(1), 9–33.

Kwon, I.W.G., & Suh, T. (2004). Factors affecting the level of trust and commitment in supply chain relationships. *Journal of Supply Chain Management*, *40*(1), 4–14.

Landry, C. (2006). *The art of city-making*. London: Routledge.
Lasker, R.D., & Weiss, E.S. (2003). Broadening participation in community problem solving: A multidisciplinary model to support collaborative practice and research. *Journal of Urban Health*, *80*(1), 14–47.
Lewis, J.M., Ricard, L.M., Klijn, E.H., & Figueras, T.Y. (2016). *Innovation in city governments: Structures, networks, and leadership*. New York: Routledge.
Liu, X., & Zheng, L. (2018). Cross-departmental collaboration in one-stop service center for smart governance in China: Factors, strategies and effectiveness. *Government Information Quarterly*, *35*(4), S54–S60.
Mattessich, P.W., Murray-Close, M., & Monsey, B.R. (2001). *Collaboration: What makes it work*. Saint Paul: Amherst H. Wilder Foundation.
McAllister, R.R., & Taylor, B.M. (2015). Partnerships for sustainability governance: A synthesis of key themes. *Current Opinion in Environmental Sustainability*, *12*, 86–90.
McNamara, M. (2012). Starting to untangle the web of cooperation, coordination, and collaboration: A framework for public managers. *International Journal of Public Administration*, *35*(6), 389–401.
Meijer, A., & Bolívar, M.P.R. (2016). Governing the smart city: A review of the literature on smart urban governance. *International Review of Administrative Sciences*, *82*(2), 392–408.
Meijer, A., & Thaens, M. (2018). Urban technological innovation: Developing and testing a sociotechnical framework for studying smart city projects. *Urban Affairs Review*, *54*(2), 363–87.
Mergel, I. (2016). Agile innovation management in government: A research agenda. *Government Information Quarterly*, *33*(3), 516–23.
Neumann, O., Matt, C., Hitz-Gamper, B.S., Schmidthuber, L., & Stürmer, M. (2019). Joining forces for public value creation? Exploring collaborative innovation in smart city initiatives. *Government Information Quarterly*, *36*(4), 101411.
O'Leary, R., & Bingham, L.B. (2007). Conclusion: Conflict and collaboration innetworks. *International Public Management Journal*, *10*(1), 103–9.
O'Leary, R., & Bingham, L.B. (Eds.) (2009). *The collaborative public manager: New ideas for the twenty-first century*. Washington, DC: Georgetown University Press.
Pereira, G.V., Cunha, M.A., Lampoltshammer, T.J., Parycek, P., & Testa, M.G. (2017). Increasing collaboration and participation in smart city governance: A cross-case analysis of smart city initiatives. *Information Technology for Development*, *23*(3), 526–53.
Pereira, G.V., Parycek, P., Falco, E., & Kleinhans, R. (2018). Smart governance in the context of smart cities: A literature review. *Information Polity*, *23*(2), 143–62.
Pfeffer, J., & Salancik, G. (1978). *The external control of organizations*. New York: Harper & Row.
Ring, P.S. & Van de Ven, A.H. (1994). Development processes of cooperative inter-organizational relationships. *Academy of Management Review*, 19, 90–118.
Ruhlandt, R.W.S. (2018). The governance of smart cities: A systematic literature review. *Cities*, *81*, 1–23.
Sancino, A., & Hudson, L. (2020). Leadership in, of, and for smart cities – case studies from Europe, America, and Australia. *Public Management Review*, *22*(5), 701–25.
Seabright, M.A., Levinthal, D.A., & Fichman, M. (1992). Role of individual attachments in the dissolution of interorganizational relationships. *The Academy of Management Journal*, *35*(1), 122–60.

Sedgwick, D. (2016). Building collaboration: Examining the relationship between collaborative processes and activities. *Journal of Public Administration Research and Theory*, *27*(2), 236–52.

Stad Antwerpen (2020). *Smart city – Convenant IMEC – Goedkeuring* (ed. P. en organisatie) (p. 16). Stad Antwerpen. https://ebesluit.antwerpen.be/publication/17.0131.4629.6049/detail (accessed 20 June 2022).

Talay, M.B., & Akdeniz, M.B. (2014). In time we trust?: The effects of duration on the dynamics of trust-building processes in inter-organizational relationships. *Strategic Management Review*, *8*(1), 77–90.

Thomson, A.M., & Perry, J.L. (2006). Collaboration processes: Inside the black box. *Public Administration Review*, *66*, 20–32.

Thomson, A.M., Perry, J.L., & Miller, T.K. (2009). Conceptualizing and measuring collaboration. *Journal of Public Administration Research and Theory*, *19*, 23–56.

Vangen, S., & Huxham, C. (2003). Nurturing collaborative relations: Building trust in interorganizational collaboration. *The Journal of Applied Behavioral Science*, *39*(1), 5–31.

Wang, C., Medaglia, R., & Zheng, L. (2018). Towards a typology of adaptive governance in the digital government context: The role of decision-making and accountability. *Government Information Quarterly*, *35*(2), 306–22.

Wang, S., & Feeney, M. (2016). Determinants of information and communication technology adoption in municipalities. *American Review of Public Administration*, *46*(3), 292–313.

Whelan, C. (2015). Managing dynamic public sector networks: Effectiveness, performance, and a methodological framework in the field of national security. *International Public Management Journal*, *18*(4), 536–67.

Wiedner, R., & Ansari, S. (2019). Collaborative uncoupling: How to break up and stay together. In Sydow, J., & Berends, H. (Eds.), *Managing inter-organizational collaborations: Process Views* (pp. 185–210). Bingley: Emerald Publishing.

Zvolska, L., Lehner, M., Voytenko Palgan, Y., Mont, O., & Plepys, A. (2019). Urban sharing in smart cities: The cases of Berlin and London. *Local Environment*, *24*(7), 628–45.

7. Leadership for intergovernmental collaboration towards digital transformation

Maike Rackwitz, Jessica Breaugh and Gerhard Hammerschmid

INTRODUCTION

Achieving public sector digital transformation and enhancing innovation raises the question of how to organise – develop, facilitate, and nurture – collaboration in and between governments (Torfing, 2012). While public administrations have long seen a fundamental shift in service delivery patterns from single bureaucracies to more complex networks across levels of government and sectors, governing such networks remains a major challenge (Ansell and Gash, 2008; Emerson, Nabatchi, and Balogh, 2012; O'Leary and Vij, 2012). This is primarily due to the necessity of balancing increased risk, power imbalances, and complexity that emerge in collaborative projects that tackle wicked and even 'tangled' issues of the 21st century (Dawes, Cresswell, and Pardo, 2009; Rackwitz et al., 2020). Adding in the element of digital-only aggravates these issues due to an increased knowledge gap in technical skills, and the need to balance the incredibly technical, yet socially dependent, components of implementing digital projects. While the tension and interaction between technical development and human action are not new (Orlikowski, Wanda, and Gash, 1994), it remains an essential, yet understudied, component in understanding how digitalisation can and should progress in the public sector from a management perspective (Carey and Wellstead, 2021). The purpose of this chapter is to examine the role of leadership in the implementation of large-scale digitalisation projects. In particular, we ask, what role does leadership play and what leadership strategies are used in the implementation of large-scale digitalisation projects?

Leadership is essential for collaboration and any type of organisational reform (Tassabehji, Hackney and Jeffares, 2016). Its function is to 'orchestrate various activities to try to overcome the various barriers to interaction, collaboration, and innovation', to greatly enhance the possibility of effective collaboration, yet must be adapted in style depending on the situation (Hartley, Sørensen, and Torfing, 2013, p. 827). It is often argued that leadership has a stronger influence on the course of partnerships than structural features such as procedural rules and regulations (Lewis et al., 2016; O'Leary and Vij, 2012). Yet, while studies on leadership in collaborative contexts are manifold, there is little empirical knowledge on the role of leadership and how it is being exercised in intergovernmental collaborations targeting digital transformation (Gasco-Hernandez, Gil-Garcia, and Luna-Reyes, 2022; Roman et al., 2019; Sancino and Hudson, 2020). As we have argued, digital transformation adds additional levels of complexity having to balance both the technical know-how of the digitalisation itself with the social and institutional components necessary to make collaboration work. As more and more digitalisation projects cross ministerial and departmental lines, this tension increases.

Collaboration always places high demands on leadership because collaboration processes are often ambiguous and uncertain (Sullivan, Williams, and Jeffares, 2012). This is exacerbated in the context of digital government transformation due to the cross-boundary nature of digitalisation projects and the need to reconcile inherently heterogeneous actors, functions, and processes. In this task, there is also a latent tension between confronting complexity and creating unity of effort (Allen, 2012), underpinned by the power and status of existing institutional practices (Pittaway and Montazemi, 2020; Tassabehji, Hackney, and Jeffares, 2016). In addition, high initial investments are typically required, and there are still limited best practices on proven procedures and project outcomes to guide them. As highlighted in previous chapters (see Chapters 2, 5, and 6), these aspects make governing intergovernmental digital projects a challenging undertaking.

Since challenges vary depending on project contexts, scope, and goals (Bryson, Crosby, and Stone, 2015), leadership approaches need to be attuned to this. However, there is scant empirical evidence surrounding what types and configurations of leadership practices are needed to best address these challenges. We attempt to address this lacuna through a review of leadership research in both the public management field and the broader management literature and then apply this knowledge to a five-country, ten-case study research project. To do so, we analyse the role of leadership in managing intergovernmental collaboration of digital projects, how leaders approach and manage the various challenges of government collaboration in the digital realm, and if and how specific types of leadership styles emerge or not.

The structure of this chapter is as follows. We first begin with a review of the literature on leadership in the context of collaboration and government and highlight key leadership approaches that are considered essential for managing collaboration in the context of digital transformation. We then introduce our empirical setting of ten collaborative digitalisation projects representing different countries and government levels and examine the extent to which the leadership approaches outlined, that is, collaborative and transactional ones, are reflected in this empirical context. The use of a cross-country case study allows us to test to a certain extent the universality of the leadership claims. We end by providing practical implications and an agenda for future research.

LEADERSHIP IN THE MANAGEMENT LITERATURE

Leadership is crucial for effective collaboration (Provan and Kenis, 2008). Some scholars even claim that leadership has a stronger influence on collaborative innovation than institutional design features such as budgets, plans, and policies (Lewis et al., 2018; O'Leary and Vij, 2012). A variety of concepts and labels for leadership has emerged in the public management literature over the years, often reflecting their zeitgeist and, consequently, what is known as 'appropriate' behaviour (Sullivan, Williams, and Jeffares, 2012). We focus on two critical approaches, collaborative and transactional leadership. Both approaches understand leadership as creating alignment of purpose and direction of the organisation as well as creating and maintaining the internal environment in which individuals can fully commit to achieving the organisation's goals (t'Hart and Tummers, 2019). Nevertheless, there are striking differences between the two. We show how in combination, they convincingly capture the intricate situation of vertical and horizontal integration of intergovernmental digitalisation projects. We then attempt to reconcile these approaches with the existing, and often disjointed, digital government literature that highlights the role of leadership in the implementation of large-scale digitalisation projects.

Collaborative Leadership Approaches

Scholars have noted that the traditional understanding of leadership may not be able to capture the complex nature of multi-actor collaboration (Huxham and Vangen, 2005), which requires a more holistic perspective on leadership to better understand the interaction between the different dynamics of collaboration, such as the people involved, the system dynamics, and the system structure itself (Morse, 2010). The specific concept of *collaborative leadership* is a means of capturing these dynamics and attempting to understand the variety of different leadership skills typically required in collaborative environments (Ansell and Gash, 2012; Hartley, Sørensen, and Torfing, 2013).

Collaborative leadership has its own understanding of who leads and how leadership is perceived. While alternative approaches describe leadership as 'making things happen' (Huxham and Vangen, 2000, p. 1161), collaborative leadership refers to empowering others by 'helping to make things happen'

(Ansell and Gash, 2008, 2012, p. 6; Sullivan, Williams, and Jeffares, 2012). This is less personalised and more role-based, which means that the task of leadership can shift, tends to fluctuate, and has a more decentralised and horizontal character (Park, 2018).

Most studies on collaborative leadership emphasise some crucial strategies that leaders need to pursue in order to facilitate collaboration such as convening, mediating, and catalysing (Ansell and Gash, 2012; Morse, 2010; Torfing, 2019).

Convening refers to bringing people together to orchestrate information sharing and develop trust by creating inclusive processes. Convening leaders act as agenda-setters, and actively engage, motivate, empower, and promote the mutual adjustment of expectations (Scott, 2011; Torfing, 2019).

Mediating focuses on developing interpersonal relationships and can arbitrate between stakeholders (Ansell and Gash, 2012). Mediating leaders seek to build trust between partners and to understand and mediate different points of view. Mediating skills also include the ability to connect different stakeholder perspectives (p. 13). In this way, leaders manage conflicts and can redefine roles and responsibilities by clarifying goals and dependencies (Crosby and Bryson, 2010). They also manage the processes by diving into different phases (Hartley, Sørensen, and Torfing, 2013). Mediating leaders also support the building of trust and relationships by understanding and managing how power and resources are distributed.

Finally, catalysing relates to creating opportunities to think outside the box by causing 'disruption' to find bold solutions to problems (Torfing, 2019, p. 7), and help '... identify and realise value-adding opportunities' (Ansell and Gash, 2012, p. 6). This involves engaging in substantive ways in negotiations to seize opportunities based on the relevant stakeholders and the big picture of outcomes (Morse, 2010), which differs from mediating because it plays a role in the solutions and benefits from the outcomes.

Based on the work of Ansell and Gash (2012) and Torfing (2019), it is clear that collaborative leadership approaches have merit in understanding how leaders should manage the cacophony of interests inherent in collaborative projects. However, the question as to if these approaches are the most appropriate for digitalisation projects now emerges. We argue that there are certainly elements that are applicable, and limitations also emerge that may require different theoretical approaches. First, digitalisation projects do indeed require the inclusion of many different stakeholders. As a result of this, collaborative leadership qualities of relationship-building emerge as highly relevant. Second, the focus on shifting leadership approaches may also emerge as an important element, especially given the variety of stakeholders who may have varying goals and interests. Finally, and relatedly, because a multiplicity of stakeholders often brings the risk of opportunism and unpredictable strategic turns, the ability to generate enthusiasm for the project and build trust is equally essential. However, while collaborative leadership provides insights into how leaders may approach the management of collaborative projects, the reality of public sector organisations, especially their bureaucratic nature, may indeed limit the application of collaborative leadership theory. The next section introduces the theory of transactional leadership, which introduces ways in which leaders can act in more pragmatic ways to push projects forward – attenuating both the complexities of organisational identities as well as the complexity of competing goals, values, and foci of these organisations.

Transactional Leadership Approaches

Although collaborative leadership is touted as the key to succeeding collaborative projects, governments still operate in bureaucratic structures marked by hierarchy, and procedural accountability remains an integral aspect of their work. This structure can have serious limitations on collaborative approaches, especially those focused on flat hierarchies and participatory structures. The hierarchy that is evident in public organisations can, in fact, be used as a means of problem-solving, especially in large, complex organisations (Meuleman, 2008). Accordingly, employing more traditional leadership skills, such as those associated with *transactional leadership*, is essential for managing projects in the context of public organisations.

In contrast to the more distributive understanding of collaborative leadership, the transactional style is characterised by single, visionary leaders. These typically take the initiative, are both authoritarian and problem-oriented, and focus on how incentive structures can be used to encourage and motivate their subordinates (Jensen et al., 2019; Lewis et al., 2018). While they have been found to be negatively related to consensus decision-making (Flood et al., 2000), similar to catalysing, transactional leaders are also described as 'negotiating agents', trained to gain power in groups (Ruggieri and Abbate, 2013, p. 1172). Their leadership approaches do not aim to 'transform' through sense-making but to develop strategies to ensure that followers comply with established rules and regulations. This is done through the principle of reward and sanction (Bass, 1985). As a result, tasks, goals, and processes are clearly defined and monitored, with associated behavioural expectations (Ruggieri and Abbate, 2013). In contrast to the 'big picture' understanding of collaborative leaders, transactional leaders define success from a comparatively short-term perspective through the acquisition of tangible achievements (Ricard et al., 2017; t'Hart and Tummers, 2019). Multiple studies have linked transactional styles to employees' well-being, performance, and goal attainment (e.g., Jensen et al., 2019), but little is known about whether this also applies to cross-boundary environments requiring both vertical and horizontal integration.

Applied to large-scale collaboration projects, the transaction leadership approaches could prove to be beneficial. This is primarily due to the pragmatic nature in which leaders sometimes must be in order to push projects out of stalemates.

Leadership for Collaborative Digital Government

In examining the literature specifically on collaboration in the digital realm, much of what the collaborative leadership literature argues is reflected in the findings of several case studies, including the essential role that leadership plays in facilitating collaborative efforts (Gasco-Hernandez, Gil-Garcia, and Luna-Reyes, 2022; Luk, 2009; McDaniel, 2005; Mergel, Edelmann, and Haug, 2019; Ruijer, 2021). Others emphasise that the alignment of resources, expertise, tasks and goals, norms and values are key to successful collaboration, that structures can be created with institutional designs, but that implementation and management can only be achieved through leadership (Picazo-Vela et al., 2018).

Despite the recognition and salience of leadership, the digital government literature does not seem to use consistent, overarching theories in relation to leadership approaches. Taking a more pragmatic approach, the majority of the research focuses on different aspects of leadership and how it is embedded within larger project dynamics and outcomes. However, their leadership references can and do align with previously established leadership theory.

For example, Chen and Lee (2018) argue that leaders need to be able to identify and close 'gaps' and manage the 'interplay' between collaborative actors, incentive structures, and institutional design. Others have noted that leaders must think long term, have broad horizons, have communication skills, be able to connect with their organisations through collaborative efforts, and '... manage across networks and leverage partnerships and resources across organisational boundaries' (Elnaghi et al., 2019, p. 198). Jones and Hooper (2017) emphasise that leaders should empower rather than dictate, while Scholl et al. (2012) highlight the importance of understanding not only the context but also the roles, responsibilities, and capabilities of project partners. In their recent work on digital champions, Wilson and Mergel (2022) further note that non-linear approaches are used by leaders in the pursuit of building a digital government. These bespoke approaches are focused on skills such as storytelling, current orientation, and on community building and validation. Concurrently, the literature calls for leadership capable of digital innovation and change to promote shared responsibility across accountability lines (Gasco-Hernandez, Gil-Garcia, and Luna-Reyes, 2022) and 'decentralized bottom-up decision-making, efforts to mobilize internal and external capabilities, wider participation to spot and internalize developments, and continuous adjustments to deal with uncertainty' (Janssen and Van der Voort, 2016, p. 4).

In contrast to these works specifically referring to theories of collaborative leadership, digital government scholars also advocate the use of agile leadership approaches, which, however, can be well associated with collaborative leadership concepts. They argue that these can be used to initially reduce risk perceptions (e.g., if there is senior/higher management support) and can lead to a change in values and organisational culture (including traditional hierarchies). Agile leaders also need to be able to lead teams in times of uncertainty, deal with experimentation, explore alternatives, and allow for failure (in conjunction with faster timelines). This is what Mergel calls 'procedural innovation', moving from contract management to a collaborative orientation (Mergel, 2016, 2018).

These aspects all draw on social and interpersonal leadership skills that reflect several of the characteristics of collaborative leadership related to catalysing, mediating, and convening as described by Ansell and Gash (2012) and Torfing (2019). Luk (2009) also refers to aspects of convening leadership styles and stresses the importance of formulating and promoting visions and strategies to facilitate the implementation of joint digital projects. However, he also stresses the importance of transactional leadership skills to understand the pressures and pulls of bureaucratic structures and a leader's ability to steer projects through them. This is consistent with the 'contingency approach' to leadership proposed by Ansell and Gash (2012).

In fact, multiple scholars, such as Clarke (2019), take a more systemic understanding of leadership and point to two different leadership models in their analysis of governance in digital government institutions. Clarke (2019) suggests the strong top-down model and the diffuse leadership model. The top-down model has features that are more akin to transactional leadership, such as the setting of standards and the power to fully direct and manage the actions of those involved in the collaboration. The diffuse leadership model, on the other hand, refers to a softer approach where units lead projects by providing more support and guidance. In this model, units are only given power and authority when asked to do so by their stakeholders.

This contingent understanding of leadership is often found when it comes to the concrete realisation and implementation of digital structures into work processes, where digital know-how becomes an asset. Besides a leader's ability to mould initial conditions in favour of collaborative change (Ruijer, 2021), the leader's technical know-how and understanding of its dimension are essential to foster better alignment between digital solutions and organisational goals (Bassellier, Benbasat, and Reich, 2001; Nograšek and Vintar, 2014). This is a focus that is labelled in the literature with terms such as IT, e- or digital leadership (Bekkers, 2012; Nograšek and Vintar, 2014; Pittaway and Montazemi, 2020; Van Wart, et al., 2019). Similarly, Pittaway and Montazemi (2020) report on implementation barriers related to IT-enabling governance structures. Once a SAP leader was added to the steering group and SAP architects were included in the teams, the problems were mitigated. Wang, Medaglia, and Zheng (2018) note that government needs to be cultivated and led by single, legitimate experts with digital knowledge, while this competency-based transactional leadership can also be brought in from external partners, such as universities, non-governmental organisations (NGOs) or companies. This knowledge must be, however, combined with strong collaborative abilities to create an increasingly seductive digital environment that favours commitment over command and tends to be highly effective despite common problems reported with IT adoption, such as 'email and data overload, worker alienation, or weak social bonding' (Roman et al., 2019, p. 855). Scholl et al. (2012) confirm the importance of project knowledge and understanding, arguing that the more complex projects are, the more skilled managers need to be.

Moreover, cross-boundary digital-related networks are likely to run into jurisdictional issues. In addition, there is tension between control and innovation that relates to the concrete concern of data misuse by partners, which reduces the incentive to share data (Klievink, Van der Voort, and Veeneman, 2018). Leadership must therefore also have the competence to manage the challenge of data ownership and control, overcoming not only social and organisational boundaries but also jurisdictional ones (Chen et al., 2019). Effective IT governance, therefore, requires respect for expertise, a commitment to resilience, performance monitoring, and sensitivity to operations (Pittaway and Montazemi, 2020). Only if government has an 'effective champion' (Chen and Ahn, 2017, p. 8) or top management support (Ziemba et al., 2016) or management with 'high power and status' (Pittaway and Montazemi, 2020, p. 6) can these requirements be met, which all underline the additional leverage of transactional leadership coupled with digital know-how.

Supporting the clout of centralised control, Luna-Reyes et al. (2016) observe that the lead organisation mainly drives technological enactment within the collaboration, ultimately determining its course. However, they also advocate for a more collaborative approach in the form of careful governance design. According to the authors, the design should empower the weaker counteracting the trend towards rulers and followers in the partnership and also revealing a gap between 'wishful thinking' and actual collaborative practice. Chen et al. (2019) and Larsson and Grönlund (2014) in turn suggest that distributing tasks and responsibilities in e-governance should, however, not be accompanied by a loss of leadership. Instead, there remains a clear need for control that ensures rules- and value-conformity based on 'human rights, equality and privacy', which proves particularly difficult given the parallel demand for horizontal and vertical distribution.

This highlights the inherent tensions that collaborations face in implementing shared digital solutions. Thus, to navigate these tensions, leadership is not about an either-or, but a multi-faceted approach that corresponds to changing and sometimes ambiguous project dynamics.

Corroborating this, Bekkers (2012), Gil-Garcia, Dawes, and Pardo (2018), and Nograšek and Vintar (2014) all believe that joint information and communication technology (ICT) integration will lead to altered forms and modes of leadership. Leaders in the digital context will have to act as 'mavericks' (Bekkers, 2012, p. 333) to effect change. This means that leaders switch between the roles of facilitator, sense-maker, and goal-setter, but also technical assistance in order to explore new solutions and push them through (Bekkers, 2012). This is similar to the concept of 'meta leadership' (Allen, 2012), which encompasses 'leading upward', 'leading downward', and 'leading across', and fits into the multi-dimensional context of cross-boundary digital government projects. By this, Allen (2012) presupposes understanding oneself and one's emotions, understanding the event or challenge correctly, leading upwards in the space between political leaders and professional or technical experts, leading downwards to support one's staff, and leading across organisational boundaries.

Against this backdrop, it can be deduced that single leadership approaches fall short of addressing joint digital transformation projects in their ambiguity and complexity. Rather, it needs a leader's ability to adapt tactics to circumstances to reflect the changing nature of collaborative projects and maintain momentum given the multi-faceted nature of projects. Therefore, combining collaborative and transactional leadership elements, complemented by relevant digital know-how, appears promising in order to analytically capture and better understand these dynamics. Table 7.1 summarises the key characteristics of collaborative, transactional, and in combination, collaborative digital government leadership. The extent to which these theoretical considerations are reflected in different practices of leadership is explored in the following section.

CASE STUDIES

Case Selection

To investigate the role of leadership in different digital government contexts and the approaches used, we compare all ten case studies of intergovernmental collaborations. These include five cases related to implementing a national single digital gateway and five cases related to implementing smart

Table 7.1 Leadership approaches derived from the literature

Leadership variable	Collaborative	Transactional	Collaborative digital government
Key characteristics	Decentralised, horizontal Convening, mediating, and catalysing Outcome-oriented Trust-building Long-term 'big picture' understanding	Single, visionary leader Strong, directive steering and controlling Problem- and rule-oriented Reward and sanction Short-term, tangible achievements	Flexible leadership combining collaborative and transactional approaches coupled with relevant digital know-how
Source	e.g., Ansell and Gash, 2008; Hartley et al., 2013; Lewis et al., 2018; Torfing, 2019	e.g., Jensen et al., 2019; Lewis et al., 2018; Ricard, et al., 2017; Ruggieri and Abbate, 2013	e.g., Clarke, 2019; Elnaghi et al., 2019; Pittaway and Montazemi, 2020; Wilson and Mergel, 2022

city strategies, showing how digital transformation manifests itself across countries and (central and local) government levels. We, therefore, draw on empirical data from five countries belonging to the five administrative traditions of Europe: Nordic (Denmark), Central and Eastern European (Estonia), Continental (Germany), Napoleonic (Belgium (mixed)), and Anglo-Saxon (United Kingdom) (Pollitt and Bouckaert, 2017). Accordingly, the leadership analysis is based on coded interviews with 64 practitioners (56 public managers, seven consultants, and one politician) who were actively involved either in one of the five national digital platforms or the five smart city cases. This gives us the widest possible range in terms of administrative tradition, level of government, and scope of challenges to work out differences and similarities regarding leadership roles or styles and to validate the findings.

FINDINGS

Overall, all cases showed that effective leadership was key to the projects' progress. An unambiguous result of the case studies is that the collaborative and transactional leadership approaches appeared far more integrated than the clear-cut examples described in the literature. The 'contingent leadership' approach was a recurring theme as leaders had to adapt to the changing circumstances of their projects. Yet, it was also evident that leaders' behaviours were somewhat constrained by the context in which they were working. Therefore, a mix of transactional and collaborative leadership styles was primarily used, which in many cases also led to increased legitimacy for the leader. In particular, key factors emerged in the projects, which include the fact that contextuality, structural levels, and the degree of perceived legitimacy determined how contingent leadership unfolds and which of the two leadership approaches ultimately gains prominence.

The emergence of leadership styles depended on a variety of different project-related factors. These include the type and scope of the projects.

The case of the Danish digital platform is a good example of how a project scope dictated the way leaders behaved. In this case, their approach was primarily characterised as transactional. This was interpreted as the most appropriate means of addressing their mandate, which was to develop a gateway and integrate it into their online service offering to provide cross-border online services. For them, the transactional approach was used because of the 'technical' nature of the project as well as its scope. It simply made more sense to engage with stakeholders individually because most of the technical components and end goals of the project (to increase EU interoperability) had already been developed based on an EU norm. This left very little collaborative space.

Overall, the transactional style of leadership was generally more common in the digital platform cases than in the smart city cases, likely due to the fact that national governments tend to be more siloed than municipal governments, and the size and scope of the projects were larger. However, the ability of leaders of the platform projects to adapt their tactics was also evident. For example, in the Belgian digital platform case, both collaborative and transactional leadership styles were used. The leaders, in this case, used a catalytic approach by organising workshops, dissemination events, and creating other opportunities for communities to voice their opinions and objections. At the same time, they used more pragmatic approaches to ensure that projects moved forward, often shifting towards a more transactional style of leadership. For example, they used their authority and power to give direction and make decisions, especially when issues were more persistent. Contingent leadership approaches further emerged and proved to be successful in managing problems and balancing different interests within projects. This includes, for example, the ability to mediate between different stakeholders, clarify decisions, and deal with challenges as they arise. While consensual approaches were mostly sought when escalation was imminent, more traditional mechanisms akin to transactional leadership approaches were used. One interviewee in the Belgian smart city case made this act of balancing and mediating very clear, they noted: 'It is my role to explain that we are sometimes not doing [these projects] to scale them up, but just to test. I then try to make sure the challenges [the city departments] face are integrated into the projects.'

In addition to mediating and intervening that emerged in the event of disagreement or rules violation, smart city project leaders saw the need to also catalyse. This was to inspire and retain project partners, especially given the voluntary nature of the networks, as one interviewee in the German smart city case emphasised: 'I consistently keep up the project's esprit so that it never threaten[ed] to become a plain administrative task, but stay[ed] an exciting, engaging project.'

Similarly, a convening style was used by the central coordinators when expectations diverged between heterogeneous project partners, as in the case of the German and the Belgian smart cities. In the Belgian smart city case, there were technical difficulties due to the low IT knowledge of the city departments and the fact that they did not know what to expect from the project. Ultimately, the coordinator clarified the process. It was, however, vital for the collaboration that the coordinator was not overly technically minded, which also contributed to his legitimacy, as one interviewee noted, putting into perspective the relevance of the leader's digital know-how. The adjustment of expectations also occurred in the platform cases but was more formalised in the form of working groups embedded in the project governing structures. These working groups brought almost all the stakeholders together, offering them the ability to become actively involved, and provided an outlet where they were able to express not only their needs and wants for the final platform but also their grievances. This was somewhat the case for every platform except for the Danish case. Interestingly, rather than lacking the IT knowledge as in the Belgian smart city case, the Estonian platform case resorted to transactional leadership in part because there we no tools or skills available to engender non-hierarchical, more collaborative means. This highlights the need to invest in collaborative competencies alongside technical infrastructures to enable appropriate choice between leadership approaches in the first place.

The type of leadership introduced depended not only on the contextual conditions and scope but also on the structural levels of the projects. For example, in the German and Belgian smart city cases, at the vertical organisational level, moving from the steering committee to the operational teams, more transactional styles were applied, with an emphasis on results and goals of the collaboration and strong, directive steering. On the horizontal level, such as at the top within the steering committee or 'in the field' within the project teams, a more profound understanding of shared responsibility in a collaborative sense was evident (see also Chapter 6). This can also be observed in the German platform case, where the basic structures were laid and communicated top-down in a rather centralised and exclusive decision-making process. By contrast, on the bilateral level or within the units subordinate to the central coordination body, such as the Secretary-General meeting, mediating leaders built trust that helped the partners 'to come out of their shells', which ultimately accelerated the process.

Another interesting finding of the case studies is the importance of the project partners' perception of the legitimacy of the project leaders. In most cases, legitimacy came from previous experience with IT projects (similar to the feedback loop mechanisms elaborated by Luna-Reyes, Gil-Garcia, and

Cruz, 2007) or from concerted efforts to build them. In most cases, simply giving a mandate was not enough. For example, an interviewee from the Estonian smart city case described the following problem: 'An external project manager was selected, but it was very difficult to give orders to people who do not obey them …'. This was also the case for the Estonian platform case, in that the project leaders already had digitalisation experience, and were therefore understood to be well respected in this area.

By contrast, a lack of legitimacy was clearly problematic and led actors to simply disengage from the collaborative process. This was particularly an issue with projects that were non-mandatory in nature. The solution to this problem was to adopt a collaborative rather than a transactional leadership style. In the case of the UK digital platform, the central coordinator was forced to change their leadership approach due to problems with participation and (lack of) legitimacy at the beginning of the project. In fact, their initial controlling approach, in the sense of transactional leadership, appeared to have led to large levels of both disinterest and distrustful project stakeholders. As a result of this, the leadership pivoted towards a more collaborative, user-centred approach, focusing on larger-scale inclusion, active community building, and trust-building among the stakeholders. To this end, the leaders created various interactive arenas to bring interested parties together. These arenas convened by collaborative leadership were critical to the success of their projects and offered a means of more informal cross-departmental communication channels. In addition, the element of time and staff rotation worked in their favour as staff moved in and out of the coordinating organisations and 'carried on' the importance of the digital effort, indicating the need for flexibility but also for distributing power and a longer-term understanding. Notwithstanding the Government Digital Services (GDS) decision to re-examine their leadership style and move towards embracing collaborative approaches, supporting our line of argumentation, they also became more pragmatic. This was most clearly seen when they began focusing on and supporting organisations that were interested in the products the GDS could offer, regardless of their size. This was a key feature of the leadership style adapting to the non-mandatory participation in the overall project.

The German, Belgian, Estonian, and Danish digital platforms as well as the German and Danish smart city had widely accepted coordinators who drew their legitimacy from their good reputation in previous projects. The Belgium and Estonian platform coordinators also had a reputation for competencies in managing large-scale projects, including those that were digital in nature. The German platform used a collaborative rather than a transactional leadership style to enforce collaborative approaches in a highly formal, dense regulatory and legal logic. According to one interviewee from this case study: '[Leaders need to] care about others, make them feel taken seriously, look after interests and bring benefits ... You have to prove that, then the process goes faster, and they come out of themselves. With violence and political attention, it is also possible, but much less sustainable.' In the German smart city case, people who had many years of experience within administrative structures were used for leadership positions. This was partly due to the lack of time for regulatory rigorous recruitment procedures but can also be understood as trust-based and distributive leadership, in the sense of a collaborative one. In some of the cases, reference was also made to 'agile leadership approaches' with clear overlaps to collaborative leadership, which is also highlighted in the literature (Mergel, 2016, 2018). For example, both the German digital platform and the Danish smart city stated that they needed to be able to deal with uncertainty, identify and close gaps, deal with experimentation and failure, and have a broad horizon to achieve their results.

CONCLUSION

This chapter aimed to decipher the role of leadership in implementing large-scale digitalisation projects within and across governments and empirically derive the various types of leadership styles used to address collaboration challenges from ten cases that differ by cultural and structural conditions.

Overall, the analysis revealed that decision-making and leadership behaviour played a crucial role in the collaborative projects. A common thread running through all the cases was the leader's ability to deal with a variety of problems, to involve stakeholders in a participatory way, and to ensure project progress. The problem-solving and strong stakeholder engagement mirrored elements of collaborative leadership, while transactional leadership emerged in the context of pragmatism and/or within the context of leaders needing to push projects forward when the projects became stalled. Through a balance of both collaborative and transactional leadership styles, leaders effectively mimicked the hybridity of structures for collaboration in and between governments. Where collaborative leadership styles were used, we noted a tendency for the catalyst to be responsible for initiating, the convener to set the agenda and sustain the project (in smart cities this was often the mayor and in the online platform cases, the project manager), and the mediator to manage the project along the way. While technical skills were indeed beneficial in some cases, people and process skills and an understanding of the 'collaborative advantage' became particularly evident. This suggests that governing novel digital intergovernmental projects does not necessarily give rise to changed forms of leadership but can be largely captured by combining existing explanatory approaches (cf. Bekkers, 2012; Nograšek and Vintar, 2014). It also underscores that transforming government through digital means is far more focused on organisational structures and processes involved in change management, rather than solely on the technical components. This further extends the theoretical models proposed by Gil-Garcia and Flores- Zúñiga (2020), showing the important role that people and people processes play across the ten case studies.

The predominance of transactional elements in platform projects at the central government level implies that collaborative forms of leadership – although deemed desirable (Luna-Reyes et al., 2016) – may not work in all contexts and, consequently, may not always represent an immediate instrumental purpose but a normative proclamation (Hammerschmid et al., 2023; Sullivan, Williams, and Jeffares, 2012). This is especially true for those with greater complexity that span multiple domains, where reaching consensus is sometimes unrealistic.

Leaders should therefore reflect on the appropriateness of their style in light of the project's goals, scope, and arenas of exchange, an aspect critical to follow-up research on collaboration. This considered, leaders need to understand themselves as vital players in the digital transformation process, able to tap into multi-faceted roles, as facilitators (stimulating cooperatives and opening up data), regulators (ensuring framework conditions), and users (deploying digital means). While this opens up unprecedented scope for design, it also necessitates careful crafting of options in regard to the range of leadership styles and their institutional and societal implications. Alongside technical infrastructures, large-scale digital projects must therefore strengthen the capacity of political and administrative leadership to develop critical collaborative competencies and networking skills (Gasco-Hernandez, Gil-Garcia, and Luna-Reyes, 2022; Roman et al., 2019; Wilson and Mergel, 2022).

Moreover, some of the cases were in their initial stage; in more mature stages, after the necessary network structures have been established, different skill sets might be required. Given that digitalisation is not a 'one-time' project, appropriate leadership approaches may also depend on the project phase. For example, in the initiation and network-building phases, interpersonal aspects in the sense of collaborative leadership may be paramount, while later more transactional styles are required, as technical scaling and compliance elements gain importance to further sustain the project flow or vice versa as illustrated by the UK platform case. Future studies examining this evolutionary aspect of project implementation are promising and will contribute to an even better understanding of how leadership can help digital transformation succeed in the long term, yielding improved service provision and public value increase.

REFERENCES

Allen, T.W. (2012). Confronting complexity and creating unity of effort: The leadership challenge for public administrators. *Public Administration Review*, *72*(3), 320–1.

Ansell, C., and Gash, A. (2008). Collaborative governance in theory and practice. *Journal of Public Administration Research and Theory*, *18*(4), 543-571.

Ansell, C., and Gash, A. (2012). Stewards, mediators, and catalysts: Toward a model of collaborative leadership. *Innovation Journal*, *17*(1), 1–21.

Bass, B.M. (1985). *Leadership and performance beyond expectations*. New York: Free Press.

Bassellier, G., Benbasat, I., and Reich, B.H. (2001). Information technology competence of business managers: A definition and result model. *Journal of Management Information Systems*, *17*, 159–82.

Bekkers, V. (2012). Why does e-government look as it does? Looking beyond the explanatory emptiness of the e-government concept. *Information Polity*, *17*(3–4), 329–42.

Bryson, J.M., Crosby, B.C., and Stone, M.M. (2015). Designing and implementing cross-sector collaborations: Needed and challenging. *Public Administration Review*, *75*(5), 647–63.

Carey, G., and Wellstead, A. (2021). Introduction: The virtual world of the public servant.*The Palgrave Handbook of the Public Servant*, 1–13. https://doi.org/10.1007/978-3-030-03008-7_109-1

Chen, Y., and Lee, J. (2018). Collaborative data networks for public service: Governance, management, and performance. *Public Management Review*, *20*(5), 672–90.

Chen, Y.C., and Ahn, M.J. (2017). The promises and opportunities of information technology in government. In *Routledge Handbook on Information Technology in Government* (pp. 3–8). Routledge.

Chen, Y.C., Hu, L.T., Tseng, K.C., Juang, W.J., and Chang, C.K. (2019). Cross-boundary e-government systems: Determinants of performance. *Government Information Quarterly*, *36*(3), 449–59.

Clarke, A. (2019). Digital government units: What are they, and what do they mean for digital era public management renewal? *International Public Management Journal*, *23*(3), 358–79.

Crosby, B., and Bryson, J. (2010). Integrative leadership and the creation and maintenance of cross-sector collaborations. *Leadership Quarterly*, *21*(2), 211–30. https://doi.org/10.1016/j.leaqua.2010.01.003

Dawes, S., Cresswell, A., and Pardo, T. (2009). From 'need to know' to 'need to share': Tangled problems, information boundaries, and the building of public sector knowledge networks. *Public Administration Review*, *69*(3), 392–402.

Elnaghi, M., Alshawi, S.N., Kamal, M.M., Weerakkody, V., and Irani, Z. (2019). Exploring the role of government authority in managing transformation in service re-engineering – experiences from Dubai police. *Government Information Quarterly*, *36*(2), 196–207.

Emerson, K., Nabatchi, T., and Balogh, S. (2012). An integrative framework for collaborative governance. *Journal of Public Administration Research and Theory*, *22*(1), 1–29.

Flood, P.C., Hannan, E., Smith, K.G., Turner, T., West, M.A., and Dawson, J. (2000). Chief executive leadership style, consensus decision making, and top management team effectiveness. *European Journal of Work and Organizational Psychology*, *9*(3), 401–20.

Gasco-Hernandez, M., Gil-Garcia, J.R., and Luna-Reyes, L.F. (2022). Unpacking the role of technology, leadership, governance and collaborative capacities in inter-agency collaborations. *Government Information Quarterly*, 101710.

Gil-Garcia, J.R., Dawes, S.S., and Pardo, T.A. (2018). Digital government and public management research: Finding the crossroads. *Public Management Review*, *20*(5), 633–46.

Gil-Garcia, J.R., and Flores-Zúñiga, M. (2020). Towards a comprehensive understanding of digital government success: Integrating implementation and adoption factors. *Government Information Quarterly*, *37*(4). https://doi.org/10.1016/J.GIQ.2020.101518

Hammerschmid, G., Palaric, E., Rackwitz, M., and Wegrich, K. (2023). A shift in paradigm? Collaborative public administration in the context of national digitalization strategies. *Governance*. https://doi.org/10.1111/gove.12778

Hartley, J., Sørensen, E., and Torfing, J. (2013). Collaborative innovation: A viable alternative to market competition and organizational entrepreneurship. *Public Administration Review*, *73*(6), 821–30.

Huxham, C., and Vangen, S. 2000. Leadership in the shaping and implementation of collaboration agendas: how things happen in a (not quite) joined-up world. *The Academy of Management Journal*, *43*(6), 1159–75.

Huxham, C., and Vangen, S. (2005) *Managing to Collaborate: The Theory and Practice of Collaborative Advantage*. London: Sage.

Janssen, M., and Van Der Voort, H. (2016). Adaptive governance: Towards a stable, accountable and responsive government. *Government Information Quarterly*, *33*(1), 1–5.

Jensen, U., Andersen, L., Bro, L., Bøllingtoft, A., Eriksen, T., Holten, A., Jacobsen, C., Ladenburg, J., Nielsen, P., Salomonsen, H., Westergård-Nielsen, N., and Würtz, A. (2019). Conceptualizing and measuring transformational and transactional leadership. *Administration and Society*, *51*(1), 3–33.

Jones, S., and Hooper, T. (2017). New Zealand's ICT strategy: The respective roles of senior and middle management in promoting collaboration and innovation. *Asia and the Pacific Policy Studies*, *4*(3), 484–95.

Klievink, B., Van Der Voort, H., and Veeneman, W. (2018). Creating value through data collaboratives. *Information Polity*, *23*(4), 379–97.

Larsson, H., and Grönlund, Å. (2014). Future-oriented eGovernance: The sustainability concept in eGov research, and ways forward. *Government Information Quarterly*, *31*(1), 137–49.

Lewis, J.M., Ricard, L.M., Klijn, E.H., and Figueras, T.Y. (2016). *Innovation in City Governments: Structures, Networks, and Leadership* (1st edn). New York: Routledge.

Lewis, J., Ricard, L., and Klijn, E. (2018). How innovation drivers, networking and leadership shape public sector innovation capacity. International Review of Administrative Sciences, 84(2), 288–307. https://doi.org/10.1177/0020852317694085

Luk, S.C.Y. (2009). The impact of leadership and stakeholders on the success/failure of e government service: Using the case study of e-stamping service in Hong Kong. *Government Information Quarterly*, *26*(4), 594–604.

Luna-Reyes, L., Gil-Garcia, J., and Cruz, C. (2007). Collaborative digital government in Mexico: Some lessons from federal Web-based interorganizational information integration initiatives. *Government Information Quarterly*, *24*(4), 808–26.

Luna-Reyes, L.F., Picazo-Vela, S., Luna, D.E., and Gil-Garcia, J.R. (2016). Creating public value through digital government: Lessons on inter-organizational collaboration and information technologies. In *2016 49th Hawaii International Conference on System Sciences (HICSS)* (pp. 2840–9). IEEE.

McDaniel, E. (2005). Facilitating cross-boundary leadership in emerging e-government leaders. *Electronic Government*, *2*(1), 1–10.
Mergel, I. (2016). Agile innovation management in government: A research agenda. *Government Information Quarterly*, *33*(3), 516–23.
Mergel, I. (2018). Open innovation in the public sector: drivers and barriers for the adoption of Challenge.gov. *Public Management Review*, *20*(5), 726–45.
Mergel, I., Edelmann, N., and Haug, N. (2019). Defining digital transformation: Results from expert interviews. *Government Information Quarterly*, *36*(4), 101385.
Meuleman, L. (2008). *Public Management and the Metagovernance of Hierarchies, Networks and Markets*. Heidelberg: Physica-Verlag HD.
Morse, R. (2010). Integrative public leadership: Catalyzing collaboration to create public value. *The Leadership Quarterly*, *21*(2), 231–45.
Nograšek, J., and Vintar, M. (2011). Technology as the key driver of organizational transformation in the egovernment period: Towards a new formal framework. In M. Janssen, H.J. Scholl, M.A. Wimmer, anf Y. Tan (Eds.), *Electronic Government: Proceedings of the 10th IFIP WG 8.5 International Conference, EGOV 2011* (Vol. 6846, pp. 453–64).
O'Leary, R., and Vij, N. (2012). Collaborative public management: Where have we been and where are we going?. *The American Review of Public Administration*, *42*(5), 507–22.
Orlikowski, Wanda J, and Gash. D. (1994). Technological frames: Making sense of information technology in organisations. *ACM Transactions on Information Systems*, *12*(2), 174–207.
Park, C.H. (2018). Cross-sector collaboration for public innovation. Journal of Public Administration Research and Theory, *28*(2), 293–5.
Picazo-Vela, S., Gutiérrez-Martínez, I., Duhamel, F., Luna, D., and Luna-Reyes, L. (2018). Value of interorganizational collaboration in digital government projects. *Public Management Review*, *20*(5), 691–708.
Pittaway, J.J., and Montazemi, A.R. (2020). Know-how to lead digital transformation: The case of local governments. *Government Information Quarterly*, *37*(4), 101474.
Pollitt, C., and Boukaert, G. (2017). *Public Management Reform: A Comparative Analysis – into the Age of Austerity*. Oxford: Oxford University Press.
Provan, K., and Kenis, P. (2008). Modes of network governance: Structure, management, and effectiveness. *Journal of Public Administration Research and Theory*, *18*(2), 229–52. https://doi.org/10.1093/jopart/mum015
Rackwitz, M., Hammerschmid, G., Breaugh, J., and Palaric, E. (2020). *Government Collaboration and Digitalisation: Comparative Case Studies on Collaboration Management for Government Digitalisation and Public Sector Innovation*. Available at: https://opus4.kobv.de/opus4-hsog/frontdoor/index/index/docId/3548, accessed 26 September 2023.
Ricard, L.M., Klijn, E.H., Lewis, J.M., and Ysa, T. (2017). Assessing public leadership styles for innovation: A comparison of Copenhagen, Rotterdam and Barcelona. *Public Management Review*, *19*(2), 134–56.
Roman, A.V., Van Wart, M., Wang, X., Liu, C., Kim, S., and McCarthy, A. (2019). Defining e-leadership as competence in ICT-mediated communications: An exploratory assessment. *Public Administration Review*, *79*(6), 853–66.
Ruijer, E. (2021). Designing and implementing data collaboratives: A governance perspective. *Government Information Quarterly*, *38*(4), 101612.

Ruggieri, S., and Abbate, C.S. (2013). Leadership style, self-sacrifice, and team identification. *Social Behavior and Personality: An International Journal*, *41*(7), 1171–8. https://doi.org/10.2224/sbp.2013.41.7.1171
Sancino, A., and Hudson, L. (2020). Leadership in, of, and for smart cities – case studies from Europe, America, and Australia. *Public Management Review*, *22*(5), 701–25.
Scholl, H., Kubicek, H., Cimander, R., and Klischewski, R. (2012). Process integration, information sharing, and system interoperation in government: A comparative case analysis. *Government Information Quarterly*, *29*(3), 313–23.
Scott, C. (2011). A case study in collaborative governance: Health care law reform in Georgia. *Conflict Resolution Quarterly*, *28*(4), 441–63.
Sullivan, H., Williams, P., and Jeffares, S. (2012). Leadership for collaboration: Situated agency in practice. *Public Management Review*, *14*(1), 41–66.
't Hart, P.T., and Tummers, L. (2019). *Understanding Public Leadership*. London: Red Globe Press.
Tassabehji, R., Hackney, R., and Popovič, A. (2016). Emergent digital era governance: Enacting the role of the 'institutional entrepreneur' in transformational change. *Government Information Quarterly*, *33*(2), 223–36.
Torfing, J. (2012). *Interactive governance: advancing the paradigm*. Oxford University Press. https://doi.org/10.1093/acprof:oso/9780199596751.001.0001
Torfing, J. (2019). Collaborative innovation in the public sector: the argument. *Public Management Review*, *21*(1), 1–11.
Van Wart, M., Roman, A., Wang, X., and Liu, C. (2019). Operationalizing the definition of e leadership: Identifying the elements of e-leadership. *International Review of Administrative Sciences*, *85*(1), 80–97.
Wang, C., Medaglia, R., and Zheng, L. (2018). Towards a typology of adaptive governance in the digital government context: The role of decision-making and accountability. *Government Information Quarterly*, *35*(2), 306–22.
Wilson, C., and Mergel, I. (2022). Overcoming barriers to digital government: apping the strategies of digital champions. *Government Information Quarterly*, *39*(2), 101681.
Ziemba, E., Papaj, T., Zelanzy, R., and Jadamus-Hacura (2016). Factors influencing the success of e-government. *Journal of Computer Information Systems*, *56*(2), 156–67.

PART III

Public-private collaboration for digital transformation and innovation[1]

8. The design and management of public-private eHealth partnerships

Chesney Callens, Koen Verhoest, Erik Hans Klijn, Lena Brogaard, Veiko Lember, Vicente Pina and Dries van Doninck

INTRODUCTION

The public sector faces an urgent need to develop new, innovative services to be able to continue delivering quality services (Sørensen and Torfing, 2011). Resource constraints, pandemics, and climate change are examples of some of the complex problems that call for innovation of public services. However, public actors cannot meet the demand to innovate on their own. Several authors from various fields propose collaboration as a mechanism for creating innovation. Arguments for innovation through collaboration can be found in the increasing complexity of industries and societies (Ketchen, Ireland, and Snow, 2007; Hartley, Sørensen, and Torfing, 2013; Crosby, 't Hart, and Torfing, 2017; Diamond and Vangen, 2017), the added value in developing something together rather than having to do it all individually (Huxham and Vangen, 2005; Leydesdorff and Cooke, 2006; Bryson, Ackerman, and Eden, 2016), the easy access to relevant knowledge and other, external sources (Chesbrough and Appleyard, 2007; Dell'Era and Verganti, 2009; Walsh, Lee, and Nagaokab, 2016), and the general fact that innovating internally is both extremely expensive and very risky (Bianchi et al., 2016; Appleyard and Chesbrough, 2017).

These incentives encourage both public and private organizations to develop innovations in collaboration with other organizations. The basic assumption is that collaboration between autonomous organizations has both a stimulating and a protective effect on innovation processes. Collaborative innovation is *stimulating* because it gives organizations access to knowledge and resources, along with all the underlying experiences, perspectives, and insights that were not previously part of the (in-house) innovation process (Davis and Eisenhardt, 2011). Collaborative innovation is *protective* because it shares the costs and risks of failure (Bruce et al., 1995; Baldwin and von Hippel, 2011; Corsaro, Cantù, and Tunisini, 2012), it shares the burden of solving complex problems (Crosby, 't Hart, and Torfing, 2017), and it creates space for experimentation that would not be tolerated in other circumstances (e.g., highly competitive conditions) (Leydesdorff, 2000; Raven et al., 2011; Markard, Raven, and Truffer, 2012; Hermans et al., 2013).

While a growing number of empirical and theoretical studies focus on innovation in public services through partnerships between public and private organizations (Brogaard, 2021; Hammond et al., 2021; Alonso and Andrews, 2022), little is known about which conditions contribute to the digital transformation of healthcare through public-private innovation. Healthcare is a core human service, where digital transformation has consequences for professionals and citizens alike. Hence, this chapter addresses the following research question: 'Under which conditions do different types of eHealth partnerships lead to innovative service delivery?' The chapter describes four clusters of conditions that may have a stimulating effect on collaborative innovation, namely, (1) the features of the partnership, (2) the features of the involved individuals and organizations, (3) the use of ICT in the collaboration, and (4) the involvement of users in the collaboration. The chapter focuses on innovation in public service delivery (specifically eHealth service delivery) and looks specifically at the conditions that affect the process of collaborative innovation (i.e., no specific attention to *ex ante* or *ex post* conditions).

DIGITAL TRANSFORMATION IN THE HEALTHCARE SECTOR

In recent years, digital transformation has become an important source for innovating services in both the public and private sectors. Digital transformation stimulates organizational efficiency and performance, but also solves complex societal issues and increases citizens' quality of life (Mergel, Edelmann, and Haug, 2019; Vial, 2019). Although digital transformation comes with some dangers, especially in terms of privacy, surveillance, security, and misuse of data (Vial, 2019), it can also have a tremendous effect on citizen's well-being and society's capability to tackle wicked problems.

The healthcare sector is a perfect example of how digital transformation can lead to innovative technologies and practices, which have a direct effect on citizen's well-being. For example, AI-based technologies that use pattern recognition software and big data to identify anomalies are already being employed to detect cancers, which has revolutionized precision oncology (Dlamini et al., 2020). Furthermore, electronic health records, which facilitate the exchange of important health data, enhance the interoperability between databases, on which a lot of digital health services ultimately depend (Kane,

2015; Lehne et al., 2019). Other health technologies are aimed at increasing the well-being of a specific audience. For instance, social robotics are being used for elderly people who suffer from loneliness (Loveys et al., 2019), but also to learn social skills for children with autism (Pennisi et al., 2016). Moreover, digital health technologies assist people in their daily activities, for instance, through telehealth and mobile health apps, devices and wearables (Steinhubl, Muse, and Topol, 2015). For this reason, eHealth is perceived as one of the key priorities of the European Union (European Commission, 2018).

However, the healthcare sector is also very complex and interconnected, in that multiple stakeholders are involved in creating, regulating, financing, and providing health services (e.g., government institutions, non-profit organizations, for-profit organizations, interest groups, and patient organizations). For this reason, digital transformation in the healthcare sector is difficult to achieve without involving a network of public and private actors. Public-private collaborations have valuable properties, which makes them well suited to pursue technological innovation. By collaborating with each other, partners learn from each other, create new ideas, share resources, and can rely on each other to implement new technologies (Sørensen and Torfing, 2011; Torfing, 2019). A recent literature review by Brogaard (2021, p. 145) confirms the importance of collaborative innovation for the healthcare sector, as the large majority of the 170 studies on public-private innovation partnerships covered by the review were conducted in healthcare-related policy fields (i.e., healthcare, eldercare, and social services).

THEORETICAL FRAMEWORK

Figure 8.1 shows our conceptual model. The research in this chapter attempts to identify four clusters of conditions in an extensive set of cases (see 'Research Design'). Before we introduce the clusters, it is important to emphasize that the innovation process itself can be divided into several phases, which are also discussed in this chapter. Different authors use different classifications of these phases, and scholars agree that these phases often overlap and are not easily delineated from each other (Meijer, 2014). However, dividing the innovation process into several phases can be useful for analytical purposes. This chapter bases the innovation process on Damanpour and Schneider's (2008) distinction between the idea generation phase and the idea adoption phase. However, we agree with other authors who include two other phases to the innovation process, one preceding the actual innovation process (i.e., the problem definition), and one connecting the idea generation phase with the idea adoption phase (i.e., the testing phase) (Rogers, 2003; Meijer, 2014). Hence, the innovation process in this chapter is composed of four phases: (1) problem definition, (2) ideation, (3) testing of ideas, and (4) adoption and implementation.

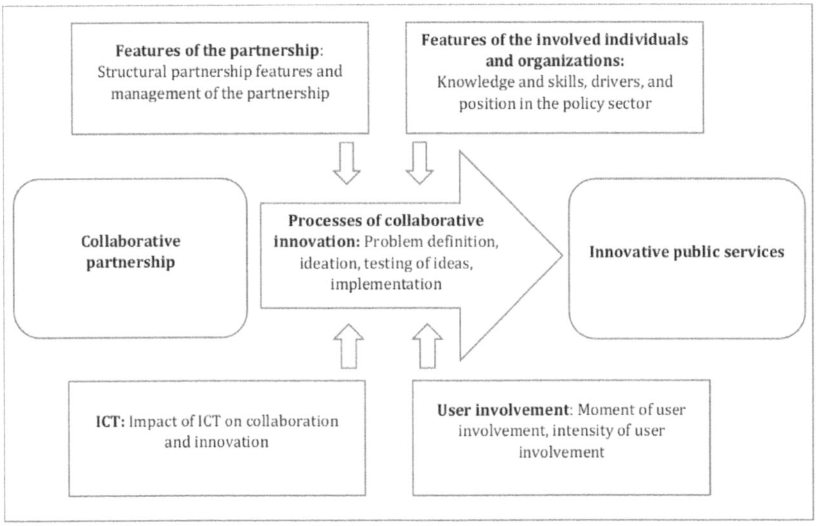

Figure 8.1 *Conceptual model*

The four clusters of conditions are developed in great detail in Chapter 4, to which we refer for a more extensive elaboration on the theoretical background of these conditions. For this chapter, we consider specific conditions from these four clusters. The first cluster of conditions includes conditions at the partnership level, namely, the *structural features* of the partnership and the used *management*. We look specifically at the size of the partnership, the governance structure of the partnership, and the contract management and network management in the partnership. For the second cluster of conditions, we consider conditions on the level of the involved *individuals and organizations*, which include conditions such as the knowledge and skills of the involved actors, their drivers to engage in the partnerships, and their position in the policy sector. The third cluster of conditions relates to the use of *information and communication technology (ICT)*. We look at two ways in which ICT can influence the collaborative innovation process, that is, through their direct impact on the collaborative dynamics in the partnership and through their impact via the broader national and regional ICT networks. The fourth cluster includes conditions related to *user involvement*, and we focus specifically on the moment of user involvement (i.e., at what stage of the innovation process) and the intensity of user involvement (e.g., informing users or co-creating with users).

RESEARCH DESIGN

An international research team from the European Horizon 2020 TROPICO[2] project collected data from a total of 19 eHealth cases in the period September 2019–February 2020: five cases in Belgium, four in the Netherlands and Spain, and three in Denmark and Estonia. We used purposeful sampling to select cases that represent public-private eHealth partnerships in Europe. Hence, the countries were selected because they represent the two major European healthcare systems (i.e., Etatist Social Health Insurance System and National Health Services, Böhm et al., 2013). In National Health Service systems, the government controls the regulation, finance, and provisioning of healthcare. In Etatist Social Health Insurance systems, regulation is conducted by the government, finance by societal actors (e.g., para-fiscal funds), and the provisioning of services by private actors (non-profit and for-profit actors). Furthermore, as each of these healthcare systems is regulated by the government, the administrative traditions of (continental) Europe were also considered in the selection of the countries (Napoleonic, Nordic, Eastern European, and Continental European, Pollitt and Bouckaert, 2017). The typology makes use of five criteria to distinguish different administrative traditions, that is, the state structure, the executive government, the minister/mandarin relations, the administrative culture, and the diversity of policy advice. Table 8.1 shows the selected cases per country.[3]

The cases were selected based on several criteria. First, all partnerships were collaborations between public actors and private actors. As these public-private collaborations can be coordinated by the public actors or the private actors, two 'types' of collaborations were selected: government-coordinated partnerships and societally coordinated partnerships. These project coordinators hold a special position in the collaboration (Klijn and Koppenjan, 2015), which is the reason why this aspect was considered during the selection of the cases. Second, all partnerships involved a third type of actor, namely, service users. These service users could be citizens and patients, but also health professionals such as GPs, nurses, specialists, etc. Third, two types of eHealth technologies were selected, which represented the most common eHealth innovations: (1) technologies that aimed to innovate the information flows between stakeholders (e.g., digital patient platforms, digital health records, integrated digital processes, etc.), and (2) technologies that aimed to innovate the end product itself (e.g., apps, smart devices, telehealth, mobile health, trackers, etc.) and distinction was made between two types of eHealth projects, namely, projects related to (1) administrative simplification and digitization of data sharing and (2) telehealth, mobile health, and smart devices. All of these innovations were

Table 8.1 *Selected cases*

Cases		Collaborative eHealth innovations
Belgium Mixed Napoleonic – Etatist Social Health Ins.	B1	National portal website which provides information for all the citizens, created through a collaboration between government agencies, ministerial cabinets, hospital networks, regional governments, private health suppliers, insurance organizations, and user organizations
	B2	Patient information sharing tool for GPs and home care organizations, created through a collaboration between private nursing organizations and federation, ministerial cabinets, national government agencies, hospital networks, individual GPs, and several private health organizations
	B3	A way of creating, validating, and disseminating official evidence-based guidelines for healthcare providers, created through a collaboration between universities, private health organizations, national and regional government agencies, red cross organizations, knowledge organizations, ICT suppliers, and individual health professionals
	B4	Several technologies in a nursing home (wearables, smart cameras, etc.), created through a collaboration between a public nursing home (local government), private construction companies and contractors, consultant companies, nurses, and patients
	B5	Platform which brings people with health/social care demands together with volunteers, created through municipalities, communal network, private hospitals, private ICT companies, consultant companies, citizens, and health professionals
The Netherlands Continental adm. retime Etatist Social Health Ins.	N1	ICT platform which facilitates the exchange of health information between partners and patients, created through the collaboration between a municipality, public hospital, and several private health organizations
	N2	Digital platform designed to foster neighbourhood collaborations between clients and consultants, created through the collaboration between a municipality, private healthcare provider, neighbourhood teams, citizens
	N3	Tracking technologies in a nursing home, created through the collaboration between a semi-private association, software developer, and patient organization
	N4	'Smart diaper' for elderly people, created through the collaboration between a semi-private association, ICT company, consultant company

Cases		Collaborative eHealth innovations
Spain Napoleonic – National Health Service	S1	Electronic prescription system, patient appointment system, robot for automatic storage and dispensing, created through the collaboration between several public hospitals, private ICT companies, several patient organizations, university
	S2	Home health ICT tools for chronic patients, created through the collaboration between a public hospital/health service, regional government, ICT companies, consultancy companies, several other private companies, universities, health professionals and patients
	S3	Web application for computerized cognitive behaviour therapy (CCBT), created through the collaboration between public hospitals and healthcare services, public research institute, private technology centre, several health professionals (e.g., psychiatrist, psychologists, physicians, etc.)
	S4	AI used to diagnose uncooperative patients, created through the collaboration between public hospitals, ICT and telecom companies, physicians
Estonia Eastern European – Etatist Social Health Ins.	E1	Centralized patient registration system, created through a collaboration between the ministry, government agencies and public authorities, ICT companies, public and private healthcare providers, physician associations, hospital associations, individual physicians
	E2	Integration of application processes for rehabilitation, disabilities, aids, created through a collaboration between ministries, public health insurance authority, government agencies, physician association, interest groups
	E3	Voice command app to guide healthcare providers, created through a collaboration between a ministry, public health insurance authority, colleges, network of healthcare providers, ICT companies, several healthcare organizations
Denmark Nordic – National Health Service	D1	E-learning programme regarding dysphagia, created through a collaboration between a regional government, municipalities, public hospitals, ICT company, representatives of health professionals
	D2	Smartphone app for patient reported outcomes, created through a collaboration between a public hospital, ICT company, health professionals
	D3	Smartphone app that helps convey the results of bone scans to patients with osteoporosis, created through a collaboration between a public hospital, university, ICT and health service companies, patient associations, health professionals

implemented, or at least tested, in the last five years before the data collection started.

For the 19 cases, data were collected through interviews and surveys. In total, data from 132 interviews and 124 surveys were collected. Data were collected from project coordinators, public partners, private partners, and users. One research team per country was responsible for collecting the data from the respective cases. To ensure a systematic data analysis, each research team used a standardized questionnaire to collect and report the interview data. Moreover, to retain the necessary contextual information, each country team wrote a detailed summary of each of the cases. These summaries helped to interpret the standardized interview data during the data analysis. To ensure a standardized data analysis, one research team was in charge of the data analysis, but the other researchers provided assistance in case some of the data were unclear or more detailed contextual data were missing. The data analysis was conducted through the use of a data matrix in Excel and NVivo in which all the qualitative data, particularly from the interviews and case study summaries, were coded for each of the conditions mentioned in the theoretical framework. Through this data matrix, a comparison between the cases became possible. Examples from the cases were extracted from the data matrix to provide evidence for the insights that were obtained from the analysis. These examples are used in the next section of the chapter.

RESULTS OF THE COMPARATIVE CASE STUDY

In the following sections, we highlight the results of the comparative case study. We discuss the case evidence that we found for the conditions in our four clusters, that is, the features of the partnership, the features of the involved individuals and organizations, the use of ICT, and the involvement of users in the collaboration. However, we first elaborate on the phases of the collaborative innovation process, to clarify how this process unfolded in the cases. In general, all cases exhibited four, often consecutive, phases: (1) Problem definition, (2) Ideation, (3) Testing of ideas, and (4) Implementation of ideas. Note that aspects of these phases can also occur in other phases. For instance, ideation is still important in the implementation phase of the innovation process, while reflecting on the adoption context (which relates to the implementation phase) is also important during the ideation phase.

Key Features of the Collaborative Innovation Process

Problem definition

Innovation processes often start with a phase of problem awareness and problem definition (Rogers, 2003), which, in collaborative innovation processes, are crucial as multiple stakeholders join the innovation process. During this phase, a clear focus on the problem should be established, which is more challenging when stakeholders from multiple backgrounds are brought together. We see this also in our comparative case study. Our cases show that a shared understanding between all the partners is important and that their perceptions regarding the problem need to be aligned. Different strategies were used by the partnerships to align the problem perceptions of the different partners. For instance, in case E1, all of the partners already recognized the scope of the problem, which made the problem definition easier. Conversely, in case S4, the partners signed a contract that explicitly described the objectives of the project, and, hence, aligned the perceptions of all of the partners regarding the problem. In case B2, the problem was raised by the service users, who had created a pilot project in which they tried to address the problem themselves. The users influenced the problem awareness and the problem definition of the project to a large extent.

Ideation

Idea exploration and generation are essential dynamics in every innovation process. In innovation partnerships, these dynamics emerge from the interaction between partners (Sørensen and Torfing, 2011, p. 842). For instance, in case D3, the partners established workshops in order to jointly develop new ideas and address the problem. This eventually led to the creation of a prototype application, from which further technological development could commence. In case S4, new ideas were developed through synergistic interaction between healthcare specialists and ICT experts. Consensus building remained the core of these ideation processes, as the partners refrained from conflicting issues and rather focused on the similarities between each other's ideas. However, the interactions between the partners during the ideation phase can also create tensions. For instance, in case D2, several partners introduced different, sometimes contradictory ideas, which caused tensions between the partners during the conceptual phases of the project. The project coordinator needed to carefully explore and align these different ideas in order to resolve these tensions and extract synergies from these interactions. Similarly, in case N2, the project coordinator made sure that the different ideas were well aligned, and that the partners saw the collaborative synergies that arose out of their interactions.

Testing of ideas

Testing new ideas in a real-life environment is an important step to identifying technological, organizational, and institutional obstacles to the implementation of innovations (Meijer, 2014, p. 202). Testing allows us to collect feedback on the innovation, and rethink, refine, and change the innovation. For instance, in case E3, the initial application was not properly matched to the working routines of the users, which became clear during the testing phase. The testing phase in case D2 uncovered some of the technological barriers to the application, as the algorithm that was being tested did not differentiate on the proper criteria, and often resulted in wrong predictions. Furthermore, in case B5, the innovators discovered during the testing phase that the use of Internet-of-Things (IoT) technologies compromised the privacy of the users, and in case E1, the testing phase promoted intensive interactions between the partners. However, testing highly advanced technologies should also be supported by proper dialogue with the users. We see in case S4, for instance, that the users were instructed on how to use the innovation and how to provide feedback on their user experience. Furthermore, in case N2, the partners established a roundtable with the involved users in order to guide and inform them properly. Note that the testing phase of the innovation process can also lead to further ideation and does not always result immediately in the implementation of the tested ideas.

Implementation of ideas

Once the innovation is ready for the adoption context, it can be implemented. However, some new barriers may arise in this stage of the innovation process, such as a lack of financial resources, implementation capabilities, or commitment from the partners (Damanpour and Schneider, 2008; Ansell and Torfing, 2014; Meijer, 2014). For instance, in case N3, the partners were highly committed to implementing the innovation, as they strongly believed in the value and benefits of the technologies for the users. In case E1, the partners had invested significant financial resources into the innovation, which allowed them to implement the innovation without any hurdles. In case B1, the partners were encouraged to commit themselves to the implementation of the innovation, as otherwise, they would fall behind their competitors in the market, who were also involved in the partnerships. Furthermore, the support of the users and other stakeholders for the implementation of the innovation proved to be crucial in several cases. For instance, in case N2, an implementation plan was introduced to provide coaching and training to the users of the innovation. In some cases, external assistance in the form of experts or consultants was brought into the project in order to properly implement the innovation. For example, in case N4, the partnerships recruited a change management consultant in order to prepare the users for the new technologies.

Conditions of Collaborative Innovation in eHealth Partnerships

Features of the partnership

Size and governance structure

Partnerships bring together actors with different knowledge, experiences, and perspectives. The synergy between these actors is a crucial advantage of collaborative innovation (Sørensen and Torfing, 2017). Yet too much diversity can also cause tensions and fragmentation. Especially in large networks, actively searching for a good balance between diversity among the partners to stimulate creative processes, on the one hand, and aligning perspectives to generate synergy, on the other, is crucial to strengthen the innovation process (Nissen, Evald, and Clarke, 2014). As Provan and Kenis (2007) argue, larger partnerships benefit from more formal and centralized types of governance structures (e.g., network-administrative organization or lead organization-governed partnerships).

Overall, we see both smaller (e.g., less than ten partners) and larger (e.g., more than ten partners) partnerships in our dataset, which largely depended on the specific task they needed to perform. In general, larger partnerships engaged in innovation processes which affected a lot of stakeholders in the healthcare sector. An example of such a partnership is case B3, which was a partnership with over 20 involved actors. The partnership aimed to counteract fragmentation in the landscape of evidence-based health professions. As a result, several other health professions were also included in the network with the intention of covering the whole field of evidence-based health practice. The large number of stakeholders and their diversity brought many different opinions and interests during the successive collaborations. Since the goal of the collaboration was to avoid fragmentation, the opinions and interests of each stakeholder had to be considered, resulting in a very complex collaboration and innovation process.

The government acted here as an important broker to align the different opinions and interests, by establishing a steering committee that centralized the decision-making authority in the network in order to better coordinate the partnership. This steering committee could push through decisions to encourage more efficient decision-making. However, tensions also arose between the core partners and the steering committee as a result of the establishment of the governance structure. Indeed, not all core partners were structurally involved in the steering committee, which led to a lack of trust by some of the core partners in the decisions of the steering committee.

However, in the majority of cases, a governance structure was not only used to manage or control the partnership but also to promote interaction among stakeholders. In case N1, a 'director's table' was created as part of the governance structure to discuss the progress of the project in the involved partner organizations. This strengthened the alignment and trust between the organizations involved. Furthermore, the involvement of key executives ensured rapid and legitimate decision-making, as they were supported by the organizations involved. In case E1, a strategic-oriented steering committee was established to supervise the actions of the project team. Stakeholders from the broader health field, such as patient representatives and health insurance funds, could be involved in the innovation process through this steering committee.

We can conclude that the size and governance structure determine and facilitate the involvement of crucial actors in the process of decision-making in the collaboration. Moreover, the governance structure specifies the different responsibilities and roles of the partners and creates supported routines for decision-making and communication between the different actors in the partnership. However, there are also some pitfalls. A lack of communication between different parts of the governance structure can result in difficult or even parallel decision-making processes and a lack of confidence in the decisions that are made.

Contract management

Formal contracts are a primary way to ensure that goals and roles are clear and aligned between partners. In our analysis, contracts were used for several reasons: to clarify interdependencies and partners' roles, to bring additional knowledge into the partnership, and to avoid conflict by clarifying accountability relationships.

Almost all projects in our cross-case analysis used written contracts. In case B2, the coordinator of the partnership emphasized the importance of the contract between the partners, calling it one of the most important incentives for collaboration. A contract was signed between several regional, but autonomous organizations and the coordinating organization (which was specifically created to connect the regional organizations and coordinate their activities). The contract was important in several ways. First, the contractual ties between the partners ensured their commitment to the project. Second, the contract also ensured the autonomy of the actors in the project, as the contract prescribed that each regional organization could decide to stop collaborating and continue working on the innovation on its own. Third, the contract also guaranteed an influx of important financial resources for the partnership to develop the innovation.

A second reason for using a contract is to bring additional knowledge that originates from an external party into the partnership. In case B4, a tender process was initiated to find a private partner to build a new residential care centre. The tender process made the expectations of the public organization clear for potential candidates. In addition, after contract closure, additional contracts were drafted that clarified the relationships between certain partners. For example, some partners contracted with each other to ensure the desired mutual accountability and prevent future disputes. In addition, the procurement contract provided a clear accountability relationship between the contractors. However, the innovation process also remained quite exploratory, partly due to the presence of a 'proof of concept' (PoC) in which different stakeholders could test the technological innovations and advise the project partners on these innovations. Thus, a clear delineation of objectives and responsibilities by using a contract does not need to preclude an exploratory innovation process, even when it legally enforces the demands of the contracting authority.

Our case studies show that contract management can effectively structure the interdependencies in a partnership. It can also clarify the roles of actors and establish accountability relationships. In addition to providing a way to coordinate the partnership, a contract can also encourage interactions. A contract can also reduce the risk of participating in the partnership. Furthermore, a contract also provides clarity on responsibilities and goals without rigidifying the innovation process. In addition, a contract can give actors autonomy and room to manoeuvre within the collaboration. Furthermore, a contract does not always hinder collaborations to explore new possibilities (e.g., through a PoC as was the case in case B4). However, the coordinators of the collaboration also need to maintain a constant balance between contract rigidity (clear goals and incentives) and contractual freedom (room to experiment and change things after the contract has been concluded), for which they might use additional process rules (see next section).

Network management

A second way to effectively manage a collaboration is to use network management strategies. In a collaboration, interactions between partners must create added value that individual partners cannot achieve as effectively or efficiently on their own. The strategies used to promote and manage interactions between actors in a collaboration are called network management strategies (Klijn, Steijn, and Edelenbos, 2010, p. 1065). Klijn, Steijn, and Edelenbos (2010) distinguish four strategies of network management. Connecting strategies focus on linking actors and resources together. Exploring strategies are aimed at fostering collaboration by searching for interdependencies, goals, and perceptions of actors. Arranging strategies focus on establishing (temporary) structures that enable interaction, consultation, and deliberation. Finally, process rules support the management of the collaboration by, for instance, implementing rules regarding the participation of new actors or the termination of the collaboration.

These network management strategies were observed in the studied cases. Some cases exhibited larger levels of these network management strategies than other cases, and not all network management strategies were used to the same degree in each case. For instance, in case D2, the coordinator was commended for facilitating the participation of actors, which can be seen as an example of a connecting strategy. By creating incentives to collaborate, the coordinator was able to activate partners and propel collaborative interactions. The result was an innovation that took the concerns and needs of all the stakeholders into consideration. Case N1 shows an example of the use of an arranging strategy. The partnership needed to meet the conditions of the funding partners (e.g., deadlines) in order to acquire their funding. To make sure that the project met the imposed deadlines, the coordinator implemented several structures, such as weekly team meetings and monthly meetings with core partners. The structured process proved to be essential to enable thorough planning of the activities of the project partners.

Second, network management also refers to the importance of resolving conflicts between cooperating partners. In this regard, contracts can also be a means of resolving conflicts. For example, in several cases, issues related to intellectual property were not resolved at an early stage, which increased the risk of intense conflicts between the partners. In cases S2 and B5, for instance, conflicts arose because potential intellectual property issues had not been raised early on. In case S2, the public partner did not want to be dependent on the private partner to further develop the application in the future. Through mediation between the public and private partners, the partners eventually agreed to a formal contract in which the intellectual property rights were settled and in which the public partner was given limited rights to further develop the application for its own use. Also, in case B5, a conflict arose between the partners because of the ambiguity of intellectual property rights. Indeed, the private partner wanted to commercialize the innovation, while the local government, where the innovation was developed and tested, and who also contributed to the ideas for the innovation, also wanted some recognition for the work they put into the project. The conflict led to a deadlock, which was only resolved by signing a contract that stipulated the intellectual property rights of both partners. Open communication between the partners regarding intellectual property rights was crucial in arriving at this contract. The case shows that conflict resolution is important to protect the innovation process from a failing collaboration. However, both the public and private partner also recognized that the discussion regarding intellectual property rights should have occurred much earlier in the project.

Note that both contract management and network management can be present in the same partnership. Indeed, whereas contract management is particularly focused on the input and output features of the collaborative innovation process (e.g., engaging innovation-oriented contractors, stimulating innovation through contract incentives and output specifications, etc.), network management is focused on the process features (e.g., exploring ideas and perspectives, connecting partners, etc.). We refer to Chapter 9 for an elaboration on the combined effect of these management practices on innovation.

Features of the Involved Individuals and Organizations

Expertise and skills of the actors

The first feature of the involved individuals and organizations in the studied partnerships concerned the expertise and skills of these actors. Four types of expertise were identified in the cases: (1) ICT expertise, (2) legal expertise, (3) medical expertise, and (4) technical expertise. ICT knowledge, which was indicated as one of the most important types of expertise and was identified in all 19 cases, was particularly related to private partners. These private partners created new ICT tools on demand from the other partners or had already created similar technologies which could be adapted to the needs of the partnership. We find an example in case D1, where the private partner involved was an authority on e-learning and also had a background in the public sector. Thanks to the private partner's experience with public actors, he spoke the same technical language as the other partners, which contributed to smooth communication with users and public representatives. Furthermore, the partner knew the hospital procedures. Legal expertise was necessary to draft contracts between the partners but was also useful in many steps of the innovation process. An example of the former is tender contracts between a public procurer and a private contractor, or contracts that formulated the arrangements on intellectual property. An example of the latter is legal knowledge on data protection, as many cases processed personal health information of patients or citizens. This type of expertise was found in seven of the cases. Medical expertise was found in nine of the cases and concerned the knowledge about medical problems (e.g., diseases) and treatments, but also about healthcare in general. This type of expertise was found a lot in the coordinators and public partners, as these actors often had a medical background. Medical knowledge was, more so than ICT knowledge, considered the backbone of the innovation process in most of cases. Technical expertise refers to knowledge about specific issues that were connected to the innovation process. For instance, in case B4, technological innovation was introduced in a new nursing home. The implementation of the technological innovation had to be aligned with the construction of the new building, for which architectural knowledge was needed.

Furthermore, important skills were introduced by the partners in the partnerships. First, network management was an important skill, which was often attributed to the project coordinator, but could also be present in other partners. As indicated before, network management refers to the improvement of interactions between actors, in order to explore differences between the actors, connect the actors, resolve conflicts, and ensure proper engagement and commitment of the involved actors. A lot of cases proved to have high levels of these network management skills. Second, project management skills were important to structure and manage the innovation process, for instance, by establishing and controlling deadlines, planning, and ensuring that all administrative requirements are met. These skills were crucial in five of the cases in which project management was often provided by the coordinator. However, in some cases, the coordinator recruited an external actor in the innovation process in order to facilitate the project management, so the coordinator could be more involved in the innovation process itself. Third, user engagement was important in all of the cases, and the skill to properly interact with users was therefore considered to be crucial in many of the cases. The partners interacted with the users by providing them with key information about the project, listening to suggestions of the users, and involving them in the conceptual and testing phases of the projects. Some partnerships even involved user engagement specialists in the partnership in order to facilitate a smooth and constructive user-partnership interaction.

Position of the actors in the policy sector

A second feature on the level of the individuals and organizations relates to their positions in the policy sector. These positions enabled some of the partners to access resources, which would otherwise have been excluded from the partnership. We already mentioned the skills and knowledge of the partners, which could be acquired through the partnerships, and which reflects their position in the policy field (e.g., ICT expertise from an ICT partner). However, some of the involved actors had a more subtle influence on the partnership through their position in the policy sector. For instance, access to relevant service users was often obtained through the involvement of specific actors (e.g., hospitals, patient organizations, etc.). This proved to be particularly important for some of the private IT partners who wanted to test the prototypes of the services they had produced, and who could not access the right service users on their own. Similarly, political support was frequently enabled by including representatives of responsible ministers or elected politicians. Moreover, in order to prevent interoperability issues between created technologies and (national) ICT infrastructures, in multiple cases the actors responsible for this infrastructure were also involved in the partnership. Even when these actors were involved at the periphery of the partnership, they often had extensive influence over the innovation process, as they were responsible for much of the data exchange infrastructure that was vital for many of the innovations.

Drivers of the actors

These different positions of the actors in the policy sector also revealed different drivers to participate in the collaborations. The first driver was related to the wish to innovate, either to reduce the costs of service delivery or to improve the quality of the services. However, not all actors were motivated by the pursuit of innovation by themselves. Some partners were involved in the innovation process in order to solve an urgent problem that directly or indirectly affected them. For instance, a lot of the hospitals were involved in the innovation project in order to find a solution for medical problems or treatments, and service users were sometimes involved because they were unable to efficiently use existing services.

A second driver came from the opportunity to develop previously created services further. Many of the private ICT partners possessed this driver, as the innovation project presented an easy way of testing their prototypes on a large scale and generalizing them to a wider audience. However, in some cases, users were already engaged in pilot projects before the innovation project was initiated. For these users, the innovation project presented a chance to attract more expertise and capacities to upscale and implement their ideas.

A third motive to participate in the innovation projects was related to the economic value of these partnerships. Private actors such as ICT companies, consultants, etc. profited directly from the innovation projects, but could also expand their market shares by tapping into new user groups. For instance, in case B5, a small start-up that was operating in a niche market was involved. The main reason for the company's involvement was not to sell or innovate their products, but to move into new markets by rigorously testing their products on a broader target group. Other types of economic incentives were also identified in the cases. For instance, a lot of the private partners were involved in the projects because the projects presented opportunities for learning and accruing new knowledge, which was of economic value for these actors. The phases of user involvement in each of the projects were particularly interesting for these actors as they enabled access to knowledge regarding user experiences, which could be used to optimize their own products and services.

Use of ICT in the collaboration

As all projects revolved around eHealth innovation, the use of ICT played a major role in almost all of the innovation projects. However, the use of ICT was not always directly related to the technological environment in which the eHealth solution was built, but also to the fact that the partners needed to work together to achieve a solution. For instance, ICT was often important in the collaboration process itself and was frequently used in the partners' interactions with the users. Through mock-ups and testing tools, the partners were able to easily involve the users in the innovation process. Through these

ICT tools, the users could test the innovations, but could also provide feedback on the innovations. For instance, a controlled testing environment was set up in cases D1, B4, and N3, which allowed the users to work with the new solutions and provide feedback on their experiences. In case B5, mock-ups of a website that allowed user feedback were used to test the final prototype of the innovation. Furthermore, a lot of communication technologies such as online interaction platforms (e.g., Skype, MS Teams, etc.) and cloud databases (e.g., Sharepoint, Dropbox, etc.) were used to connect disparate partners together and coordinate their work. However, the analytical capabilities of some technologies were also important to support decision-making. For instance, in case B5, the private partner used a software tool to visualize and analyze the desired process flow of the solution. In case D3, the coordinator was granted access to the back end of the solution, in order to thoroughly test the solution. This also allowed the coordinator to directly add new content to the solution without always needing the developers, which increased the efficiency of the innovation process.

The pre-existing ICT infrastructure played a second major role in the studied projects, especially in the projects which were aimed at innovating the digital information flows between stakeholders (e.g., national health platforms). ICT infrastructure such as eHealth networks enabled a lot of the initiatives of the partnerships. For instance, cases B1 and B2 were highly dependent on the Belgian eHealth platform for their success, as did case E2, which depended on the Estonian X-road. These eHealth networks facilitated the access and exchange of crucial citizen and patient information, which was the backbone of many of these innovations. For instance, the use of the Belgian eHealth platform made it possible for case B2 to connect to other eHealth databases, but at the same time, its own databases became connected to other health actors, which significantly enlarged the impact of its innovation. Hence, the innovation became part of the ecosystem of eHealth services. However, the existing ICT infrastructure might also pose new challenges for the innovating partnership, particularly in terms of interoperability. For instance, the project in case S3 was significantly delayed because of the incompatibility of the innovation with the existing ICT infrastructure. Moreover, the technical characteristics of the ICT infrastructure might influence the design of the innovation. In cases E1 and E2, the innovation needed to use the data formats of the X-road, which reduced the creative freedom of the developers and meant that the innovation needed to be designed with the X-road in mind. We see something similar in case B1, in which the original idea was to build a personal health record, which centralizes all the available, digitalized patient information. However, due to the already existing network of hospital hubs (i.e., networks established around the major hospitals), the idea of a central health record was abandoned, and a health portal was created instead.

The involvement of users in the collaboration

Timing of user involvement
The first dimension that was important in the case studies is the timing of user involvement. If users are involved early on in the innovation process, there are more opportunities to integrate their input into the final innovation, which we saw in most of the partnerships which created highly innovative services. In addition, users' views provide information that is best included already when generating innovative ideas. In this way, ideas are generated that are directly grafted onto users' concerns. For instance, in case B2, a pilot project was initiated before the start of the actual project by a group of users (general practitioners), and these users were later also included in the project itself. The involvement of the users and the experiences from the pilot project both accrued knowledge that was useful for idea generation. In case D1, users were involved early on in the problem definition phase. A survey of users was conducted to further identify the main problems surrounding the treatment of dysphagia. After the survey, seminars were organized to invite private partners to come up with ideas. By involving users at the problem definition stage, stakeholders were also highly motivated to implement the innovation.

Intensity of user involvement
A second aspect of successful user engagement is related to the intensity with which the users are engaged, which refers to the extent to which the input of users is considered in decision-making processes. Intensive involvement of the users was particularly important in partnerships which were dependent on the users to make decisions on the content and development of the innovation. This could be both in conceptual phases of the innovation process and in testing and implementation phases. In order to develop user-centred innovations, it is not enough to merely inform users. User input must actually be included in decision-making or users must be given real decision-making power. In case B5, for example, the whole concept of the innovation changed when users were involved in the innovation process. Initially, the idea was to develop an IoT solution. However, by involving the users, it quickly became clear that the users were not entirely comfortable with this. Despite the promise to the subsidizing government, an IoT solution was abandoned in favour of less intrusive telephone technology. The innovation was ultimately well received by users.

Another example is that of case N2, where user involvement was tightly organized. The project team employed a strict protocol with instructions on how to give feedback to the ICT partner. Some respondents pointed to the lack of openness in this process of user involvement. By setting too strict conditions for user involvement, users were not always able to openly express their preferences and opinions to the ICT partner. In case N4, the project team anticipated such problems and built the inclusion of user feedback into the innovation process. The project team ensured that enough time was left to incorporate user feedback into the application and for trial and error to optimize the innovation. This was necessary because users were not involved in the design process and several shortcomings in the design (related to the comfort of the application for patients and the effectiveness of the device) were raised by users after the design process.

CONCLUSION

The comparative case study in this chapter shows that the collaborative innovation process in eHealth partnerships is subject to a large variety of conditions related to the design and management of the partnership. From this research, we learn that including the right number of relevant actors in the partnership in combination with introducing a governance structure that allows effective management of the partnership is crucial for successful collaborative innovation processes. We also illustrated the impact of direct management activities on the collaborative relationships in our cases. Contract management provides a design framework in which the objectives of the partners are aligned, responsibilities are secured and enforced, interactions amongst the partners are encouraged, and risks between the partners are reduced. Network management directly influences the interactions between the partners, encourages learning, allows innovative ideas to emerge, and remedies tensions and conflicts. Furthermore, successful partnerships include the relevant resources (mostly in the form of expertise and skills of the involved partners, and the use of ICT) and influence, by engaging partners from a multitude of different backgrounds and positions in the healthcare sector and involving service users in different phases of the innovation process.

A synthesis of the results points towards two interrelated spectra over which the conditions push and pull the collaborative innovation process. The first spectrum corresponds to the degree of collaborative stability. The higher the partnership scores on this spectrum, the more likely that the collaboration can be maintained and that the partners will jointly develop a certain outcome. Conditions such as the presence or absence of an adequate balance between size and governance structure, contract management and particular network management strategies (e.g., arranging strategies and process rules), ICT to enhance the collaboration, and influential stakeholders enable the partnership to move alongside this spectrum. A second spectrum relates to the degree of innovative impetus. The higher the partnership scores on this spectrum, the more it is motivated by the prospect of developing a creative and innovative solution. Conditions such as particular network management strategies (e.g., exploring and connecting), the involvement of a broad range of stakeholders that unite a lot of expertise and skills, and the involvement of users through which new knowledge can be accrued, push the collaborative innovation process higher up this spectrum.

Although the premise should be to elevate the partnership alongside both spectra, literature and practice show that this is quite challenging as encouraging collaborative stability might interfere with achieving innovative impetus and vice versa. For instance, focusing managerial attention on contract management might indeed make the partnership more stable, but it might also extinguish creative experimentation and trial-and-error behaviour because the contract conditions limit flexibility in the development process. On the other hand, stimulating innovative impetus by involving a multitude of different actors, including a lot of service users, and letting them freely engage and interact with each other might lead to interpersonal conflicts and a disintegration of joint objectives. The goal should therefore be to achieve a 'desirable' balance between collaborative stability and innovative impetus. Our comparative case study does not provide answers to what this desirable balance is, and more detailed research is needed into the combined effects of particular conditions to provide these answers. We start this endeavour in Chapter 9, in which we consider the combined effect of contract management and network management on the innovativeness of eHealth services. Nevertheless, future research should investigate how the other conditions can be optimally configured in order to enhance the collaborative innovation process.

NOTES

1. This part of the chapter is based on the results of the comparative case study, reported by Callens et al. (2020).
2. The TROPICO project received funding from the European Union's Horizon 2020 research and innovation programme under grant agreement No. 726840. This article reflects only the author's view and the Research Executive Agency of the European Commission is not responsible for any use that may be made of the information that the article contains. For more information: https://cordis.europa.eu/project/id/726840.
3. More detailed case information can be found in the TROPICO case study repository: https://tropico-project.eu/case-studies/

REFERENCES

Alonso, J.M., and Andrews, R. (2022). Can public-private innovation partnerships improve public services? Evidence from a synthetic control approach. *Public Administration Review*, *82*(6), 1138–53.

Ansell, C., and Torfing, J. (2014). *Public Innovation through Collaboration and Design*. New York: Routledge.

Appleyard, M., and Chesbrough, H.W. (2017). The dynamics of open strategy: From adoption to reversion. *Long Range Planning*, *50*(3), 310–21.
Baldwin, C., and von Hippel, E. (2011). Modeling a Paradigm shift: From producer innovation to user and open collaborative innovation. *Organization Science*, *22*(6), 1399–417.
Bianchi, M., Croce, A., Dell'Era, C., Di Benedetto, C.A., and Frattini, F. (2016). Organizing for inbound open innovation: How external consultants and a dedicated R&D unit influence product innovation performance. *Journal of Product Innovation Management*, *33*(4), 492–510.
Böhm, K., Schmid, A., Götze, R., Landwehr, C., and Rothgang, H. (2013). Five types of OECD healthcare systems: Empirical results of a deductive classification. *Health Policy*, *113*(3), 258–69.
Brogaard, L. (2021). Innovative outcomes in public-private innovation partnerships: A systematic review of empirical evidence and current challenges. *Public Management Review*, *23*(1), 135–57.
Bruce, M., Leverick, F., Littler, D., and Wilson, D. (1995). Success factors for collaborative product development: A study of suppliers of information and communication technology. *R&D Management*, *25*(1), 33–44.
Bryson, J.M., Ackermann, F., and Eden, C. (2016). Discovering collaborative advantage: The contributions of goal categories and visual strategy mapping. *Public Administration Review*, *76*(6), 912–25.
Callens, C., Verhoest, K., van Doninck, D., and Dockx, E. (2020). Practices of external collaboration for service delivery – comparative case studies on external collaboration in eHealth partnerships. Report European Commission under the European Union's Horizon 2020 research and innovation programme, grant agreement no. 726840, https://tropico-project.eu/download/d7-1-practices-of-external-collaboration-for-service-delivery-comparative-case-studies-on-external-collaboration-in-ehealth-partnerships/?wpdmdl=1743andrefresh=615d51089f8681633505544
Chesbrough, H.W., and Appleyard, M M. (2007). Open innovation and strategy. *California Management Review*, *50*(1), 57–76.
Corsaro, D., Cantù, C., and Tunisini, A. (2012). Actors' heterogeneity in innovation networks. *Industrial Marketing Management*, *41*(5), 780–9.
Crosby, B.C., 't Hart, P., and Torfing, J. (2017). Public value creation through collaborative innovation. *Public Management Review*, *19*(5), 655–69.
Damanpour, F., and Schneider, M. (2008). Characteristics of innovation and innovation adoption in public organizations: Assessing the role of managers. *Journal of Public Administration Research and Theory*, *19*(3), 495–522.
Davis, J.P., and Eisenhardt, K.M. (2011). Rotating leadership and collaborative innovation: Recombination processes in symbiotic relationships. *Administrative Science Quarterly*, *56*, 159–201.
Dell'Era, C., and Verganti, R. (2009). Design-driven laboratories: Organization and strategy of laboratories specialized in the development of radical design-driven innovations. *R&D Management*, *39*(1), 1–20.
Diamond, J., and Vangen, S. (2017). Coping with austerity: Innovation via collaboration or retreat to the known? *Public Money and Management*, *37*(1), 47–54.
Dlamini, Z., Francies, F.Z., Hull, R., and Marima, R. (2020). Artificial intelligence (AI) and big data in cancer and precision oncology. *Computational and Structural Biotechnology Journal*, *18*, 2300–11.
European Commission. (2018). Communication from the Commission to the European Parliament, the Council, the European Economic and Social Committee and the

Committee of the Regions on enabling the digital transformation of health and care in the Digital Single Market; empowering citizens and building a healthier society (25/4/2018).

Hammond, J., Bailey, S., Gore, O., Checkland, K., Darley, S., McDonald, R., and Blakeman, T. (2021). The problem of success and failure in public-private innovation partnerships. *Journal of Social Policy*, *51*(4), 1–21.

Hartley, J., Sørensen, E. and Torfing, J. (2013). Collaborative Innovation: A Viable Alternative to Market Competition and Organizational Entrepreneurship. *Public Administration Review*, *73*(6), 821–830.

Hermans, F., van Apeldoorn, D., Stuiverd, M., and Kok, K. (2013). Niches and networks: Explaining network evolution through niche formation processes. *Research Policy*, *42*(3), 613-623.

Huxham, C., and Vangen, S. (2005). *Managing to Collaborate: The Theory and Practice of Collaborative Advantage*. Abingdon, UK: Routledge.

Kane, G.C. 2015. How digital transformation is making health care safer, faster and cheaper. *MIT Sloan Management Review*, *57*(1), 1–11.

Ketchen, D.J. Jr, Ireland, R.D., Snow, C.C. (2007). Strategic entrepreneurship, collaborative innovation, and wealth creation. *Strategic Entrepreneurship Journal*, *1*, 371–85.

Klijn, E., and Koppenjan, J. (2015). *Governance Networks in the Public Sector*. London: Routledge.

Klijn, E., Steijn, B., and Edelenbos, J. (2010). The impact of network management on outcomes in governance networks. *Public Administration*, *88*(4), 1063–82.

Lehne, M., Sass, J., Essenwanger, A., Schepers, J., and Thun, S. (2019). Why digital medicine depends on interoperability. *Digital Medicine*, *79*, 1–5.

Leydesdorff, L. (2000). The triple helix: An evolutionary model of innovations. *Research Policy*, *29*(2), 243–55.

Leydesdorff, L., and Cooke, P. (2006). Regional development in the knowledge-based economy: The construction of advantage. *Journal of Technology Transfer*, *31*, 5–15.

Loveys, K., Fricchione, G., Kolappa, K., Sagar, M., and Broadbent, E. (2019). Reducing patient loneliness with artificial agents: Design insights from evolutionary neuropsychiatry. *Journal of Medical Internet Research*, *21*(7), doi:10.2196/13664.

Markard, J., Raven, R., and Truffer, B. (2012). Sustainability transitions: An emerging field of research and its prospects. *Research Policy*, *41*(6), 955–67.

Meijer, A.J. (2014). From hero-innovators to distributed heroism: An in-depth analysis of the role of individuals in public sector innovation. *Public Management Review*, *16*(2), 199–216.

Mergel, I., Edelmann, N., and Haug, N. (2019). Defining digital transformation: Results from expert interviews. *Government Information Quarterly*, *36*(4), 101385.

Nissen, H.A., Evald, M.R., and Clarke, A.H. (2014). Knowledge sharing in heterogeneous teams through collaboration and cooperation: Exemplified through public–private-innovation partnerships. *Industrial Marketing Management*, *43*(3), 473–82.

Pennisi, P., Tonacci, A., Tartarisco, G., Billeci, L., Ruta, L., Gangemi, S., and Pioggia, G. (2016). Autism and social robotics: A systematic review. *Autism Research*, *9*(2), 165–83.

Pollitt, C., and Bouckaert, G. (2017). *Public Management Reform: A Comparative Analysis – into the Age of Austerity*. Oxford: Oxford University Press.

Provan, K., and Kenis, P. (2007). Modes of network governance: Structure, management and effectiveness. *Journal of Public Administration Research and Theory*, *18*(2), 229–52.

Raven, R.P.J.M., Verbong, G.P.J., Schilpzand, W.F., and Witkamp, M.J. (2011). Translation mechanisms in socio-technical niches: A case study of Dutch river management. *Technology Analysis and Strategic Management*, *23*(10), 1063–78.
Rogers, E. (2003). *Diffusion of Innovations* (5th edn). New York: Free Press.
Sørensen, E. and Torfing, J. (2011). Enhancing collaborative innovation in the public sector. *Administration and Society*, *43*(8), 842–68.
Sørensen, E., and Torfing, J. (2017). Metagoverning collaborative innovation in governance networks. *American Review of Public Administration*, *47*(7), 826–39.
Steinhubl, S.R., Muse, E.D., and Topol, E.J. (2015). The emerging field of mobile health. *Science Translational Medicine*, *7*(283), doi: 10.1126/scitranslmed.aaa3487.
Torfing, J. (2019). Collaborative innovation in the public sector: The argument. *Public Management Review*, *21*(1), 1–11.
Vial, G. (2019). Understanding digital transformation: A review and a research agenda. *Journal of Strategic Information Systems*, *28*(2), 118–44.
Walsh, J.P., Lee, Y.-N., and Nagaoka, S. (2016). Openness and innovation in the US: Collaboration form, idea generation and implementation. *Research Policy*, *45*(8), 1660–71.

9. Contract management and network management in public-private eHealth partnerships

Chesney Callens and Erik Hans Klijn

INTRODUCTION

The healthcare sector is a complex and turbulent environment in which multiple stakeholders are interconnected and demands for new services can quickly emerge. Digital transformation through technological innovation can give healthcare stakeholders the tools to answer these new demands and solve complicated problems that have no obvious solutions (Mergel et al., 2019). Through digital transformation, healthcare information can be exchanged more quickly and more accurately, medical procedures can be made more efficient and less risk-prone, patients can be assisted better in their recovery process, and elderly people can be supported to make their lives more comfortable. As Callens et al. mention in Chapter 8, in the last decade, several of these eHealth technologies have emerged, going from electronic health records, which allow the exchange of patient information between governments and healthcare providers (Kane, 2015; Lehne et al., 2019), to mobile health apps, devices, and robotics, which assist people in their daily activities (Loveys et al., 2019; Pennisi et al., 2016; Steinhubl et al., 2015).

However, digital transformation in the healthcare sector is difficult for individual organizations because of its interconnectedness and complex nature. In a recent literature review on digital transformation in the healthcare sector, Kraus et al. (2021) argue that the contemporary healthcare sector actually consists of healthcare ecosystems, in which various stakeholders together exploit digital technologies to increase the quality of healthcare services. For instance, public hospitals depend on digital patient information from government institutions, pharmacies, general practitioners, nursing homes, home care organizations, etc. Furthermore, digital transformation in the healthcare sector often requires specific technological and organizational knowledge from experts such as information and communication technology (ICT) experts and business consultants. Moreover, as eHealth innovation creates new services for users, these users are usually at the centre of the innovation process.

This complicated network of stakeholders requires innovators to look beyond the boundaries of their own organizations or institutions and collaborate with the network stakeholders to innovate their services. This type of innovation, that is, collaborative innovation (Sørensen and Torfing, 2011), has the potential to increase the innovativeness of the created technologies for three reasons. First, collaborative innovation increases creative idea generation, as multiple actors from different backgrounds join the innovation process (Torfing, 2019). Hence, new knowledge pools are accessed, out of which creative synergies and enhanced problem-solving capabilities can arise. Second, by collaborating with the network stakeholders, the quality of the evaluation and selection of ideas is improved, as all relevant actors are part of the innovation process, and the diversity of the involved actors decreases the likelihood of groupthink (Hale and Woronkowicz, 2021). Third, collaboration facilitates the implementation of innovation as broader support is achieved by connecting the relevant stakeholders, and the involved actors can share their capabilities to implement the innovation (Torfing et al., 2020).

This chapter addresses how partnerships between public and private stakeholders are managed to create technological innovation. We work further on the results of the explorative analyses of Chapter 8, in which it became clear that specific conditions of contract management and network management were important to produce innovation in the studied eHealth partnership. We see this reflected in the literature on projects, public-private partnerships, and governance, in which there is a lot of attention on the question of how to manage such partnerships (Warsen et al., 2019; Callens et al., 2022). These streams of literature often have quite different perspectives about the best way to manage these partnerships.

In general, two large management traditions can be identified. The first can be found in the more economic-oriented literature. This tradition emphasizes that these projects are dominated by contracts which should deal with the possible opportunistic behaviour of actors (Jensen and Meckling, 1976). Contracts, roughly speaking, then provide the means (sanction and performance indicators for monitoring) to keep the project in line, that is, to secure on time and on budget delivery (Savas, 2000). Thus, contract management aims at monitoring performance and using the contract as a possible 'stick' in case performance criteria are not met. Innovation must be achieved by clear performance criteria related to the desired innovation and innovation standards.

The other tradition emphasizes that although these projects are dominated by contracts, they are also collaborative processes in networks of actors around projects. Such projects are often characterized by their complex dynamics. Nothing happens entirely according to plan, so there is a constant need to adjust the content of the project and invest in the collaborative process that is taking place to achieve good and certainly innovative results (Agranoff and McGuire, 2003; Huxham and Vangen, 2005; Steijn et al., 2011; Emerson and Nabatchi, 2015). This tradition thus tells us that in these projects it is impossible to foresee all unexpected events and developments and coordinate the behaviour of the partners. In this reasoning, managing the daily interactions and relations between partners in the network, called network management, is essential for achieving good outcomes. In this tradition, collaboration, exchange of information, and sharing of knowledge and skills are essential for innovation (Torfing, 2019).

So far, we have seen some empirical evidence mainly done on (large) environmental or infrastructural projects (e.g., Warsen et al., 2019, Callens et al., 2022) that show that public-private partnerships (PPPs) benefit from both contract-oriented managerial incentives combined with more governance-oriented strategies. However, there are few studies that look at other fields.

In this chapter, we take a closer look at the case study data from Chapter 8, in which 19 different eHealth projects in five different countries were studied. Using this dataset, we explore which of the two mentioned management traditions deliver the most promising management practices for innovation partnerships and argue that we actually need both of these traditions to create highly innovative services in those partnerships. To do so, we first provide a theoretical framework that captures the two traditions and shows how the management practices related to those traditions can be combined to produce collaborative innovation. Next, we perform a fuzzy-set qualitative comparative analysis (fsQCA) on the 19 eHealth partnerships to determine the relationship between this combination of management practices and innovation. We conclude this chapter by discussing these results and by providing management recommendations for practitioners.

THEORETICAL PERSPECTIVES: INFLUENCING INNOVATION THROUGH CONTRACT MANAGEMENT AND NETWORK MANAGEMENT

The two management traditions mentioned in the introduction each have their theoretical backgrounds. However, before we introduce these two traditions and their respective practices, we need to properly introduce what we mean by innovation, as the concept is often vaguely defined (Torfing, 2019).

Innovation

In the literature, a wide range of criteria is mentioned for evaluating public-private collaborations, of which performance and innovation are the most common. Performance is usually measured by evaluation criteria, such as whether the project stays within the budget (Mantel, 2005), the degree to which cost overruns are made (Flyvberg et al., 2003), and the balance between benefits and costs (Mantel, 2005). Innovation of the outcomes is a bit more difficult to measure than performance in terms of costs and budget. In general, innovation is characterized by two features. First, an innovation is considered as an 'idea, practice, or object that is perceived as new by an individual or other unit of adoption' (Rogers, 2003, p.12). The perceived newness of an innovation is therefore an essential component of the concept. Second, new ideas can only be regarded as innovations if they are adopted in a real-life context (Anderson et al., 2004; Walker, 2007). This aspect of innovation distinguishes it from related concepts such as creativity and invention (Anderson et al., 2014).

Furthermore, innovation can both refer to the product itself (product innovation) or to the process in which the product is realized and used (Nooteboom, 2002; Torfing, 2019). The eHealth collaborations we studied in this chapter include both product innovations such as telehealth and mobile health tools, robotics, wearables, etc., and process innovations such as new ways to exchange patient information, data centralization tools, central communication, monitoring system, etc.

Contract Management: The Importance of Innovation Output Criteria

Public-private collaborations are often perceived as instruments of public procurement, as a public procurer collaborates with a private contractor to procure a product or service (Grimsey and Lewis, 2007). This view on public-private collaboration is particularly common in literature on PPPs. In this literature, innovation is achieved through a demand-sided rationale, in which a procurer demands innovation from a contractor through contractual means (Callens et al., 2022). Such a 'procurement for innovation' relies heavily on contract conditions to manage the collaboration and to control the output of the collaboration (Edquist et al., 2015).

Strict contract management is an essential aspect of procurement for innovation because it secures both 'compliance' of the (private) contract party that realizes the service and specifies the indicators for the desired innovation. In this reasoning, strongly embraced in transaction economics (Jensen and Mecklin, 1976; Williamson, 1996; Brown et al., 2016), principal-agent literature, and (more classical) project management literature (Mantel, 2005), the content of projects and services are fixed by clear indicators and specification of the product. Success in implementation depends on this specification and the performance indicators which derive from it.

It is important in this way of looking at public-private collaborations that they are seen as relations between a principal (usually the public actor that initiates and commissions the project/service) and an agent (the private consortium that realizes the product/service). This relation is characterized first of all by opportunistic behaviour and necessary (incomplete) information collection (Akintoye et al., 2008; Savas, 2000). The principal must know how the agent performs but, for making that judgement, the principal needs information about the agent's behaviour to monitor the agent and its performance. The agent, however, has a better-informed position about her own actions and overall performance and will try to hide that information. After all, keeping that information for itself as much as possible provides the agent with a better position and more possibilities to underperform.

So, monitoring is crucial because the implicit assumption is that the process will be dominated by the opportunistic behaviour of the implementing party (Williamson, 1996). That party will try to put less work into the project than agreed to, or otherwise try to maximize profit at the cost of the contractor (Jensen and Mecklin, 1976). This means clear specifications of the innovation upfront is important and help to guide the project in its process (Mantel, 2005). Thus, strict contract management for the purpose of procurement for innovation is mainly setting good and measurable innovation indicators and monitoring them.

Uyarra et al. (2014) synthesize these contract management practices, of which two of the most often used are 'output specifications' and 'contractual incentives'. *Output specifications* are detailed specifications in a tender or contract regarding the features of the delivered innovation. Such output specifications are preferred to detailed project designs which might hinder contractors to propose innovative solutions (Geroski, 1990). *Contractual incentives* to innovate refer to the criteria the procurer uses to stimulate the contractors to work towards a desirable, innovative solution (Georghiou et al., 2014). Indeed, even when output specifications regarding the innovation are stipulated in a contract, the contractor might still not be incentivized to innovate and reuse already existing solutions (Uyarra et al., 2014). Contractual incentives to innovate emphasize that ithe nnovative behaviour of the contractor (e.g., through exploring and pooling new knowledge, experimentation, and testing) is required.

Network Management: Connecting and Exploring in Complex Processes

Against the transaction economic perspectives elaborated above, one could position a quite different perspective on public-private collaboration which comes from the extensive literature on collaborative and network governance of the last decades (Kickert et al., 1997; Huxham and Vangen, 2005; Ansell and Gash, 2008; Emerson and Nabatchi, 2015; Klijn and Koppenjan, 2016). In this perspective, the complexity of these partnerships is emphasized. It shares the idea with the economic perspective that actors have different interests and conflicts will emerge during the project. However, it also stresses both the interdependency between the actors and the fact that interactions between partners are dynamic and characterized by a lot of unforeseen events and developments. This literature argues that these complexities cannot be regulated by a contract because contracts are by nature incomplete and cannot foresee future events and developments. The different strategic actions of the partners alone, but also external events that happen will certainly affect the partnership (Huxham and Vangen, 2005; Ansell and Gash, 2008). Thus, active management, usually called network management, is needed to achieve good results in partnerships (Steijn et al., 2011; Callens et al., 2022).

Collaborative and network governance literature stresses that it is highly unlikely that contracts and monitoring are sufficient to deal with uncertainty and changing events (Huxham and Vangen, 2005). When innovative solutions are needed, this argument becomes even more important. Technological innovations are the output of innovation processes which combine an idea generation phase with a phase of idea implementation (Damanpour and Schneider, 2008). In both phases of the innovation process, a lot of information exchange and coordination between the partners is needed. Ideas and perspectives of the involved actors need to be explored when developing novel ideas, and partners and their resources need to be sufficiently connected to select, test, and implement these innovative ideas (Callens et al., 2022).

The literature mentions a wide variety of network management strategies to guide and structure interaction processes, so an exhaustive list is difficult to provide (Gage and Mandell, 1990; Agranoff and McGuire, 2003). Table 9.1 provides a summary (of the types of strategies that have been identified, providing examples of each of the categories (Klijn et al., 2010). In the following, we shortly discuss the various types of network management strategies.

In networks, many actors with various organizational backgrounds are active and need to be *connected*. Network managers thus act as in-between actors as they try to establish connections among various actors and other project activities in the network (Van Meerkerk and Edelenbos, 2014). The network management literature emphasizes that the network manager first needs to identify the actors required for an initiative and actually create a situation in which they become interested in investing their resources (on activation, see Scharpf, 1978; Klijn and Koppenjan, 2016). The interactions in the collaborative process itself also have to be managed. This can be done

Table 9.1 *Overview of network management strategies*

Types of strategies	Process agreements	Exploring content	Arranging	Connecting
Main strategies mentioned in the literature	Rules for entrance into or exit from the process, conflict regulating rules, rules that specify the interests of actors or veto possibilities, rules that inform actors about the availability of information about decision-making moments, etc.	Searching for goal congruency, creating variation in solutions, influencing (and explicating) perceptions, managing and collecting information and research, creating variation through creative competition	Creating new ad hoc organizational arrangements (boards, project organizations, etc.)	Selective (de) activation of actors, resource mobilizing, initiating new series of interactions, coalition building, mediation, appointment of process managers, removing obstacles to cooperation, creating incentives for cooperation

Source: Adapted from Klijn et al. (2010).

by appointing a process manager or broker, who invests time and energy in connecting the actions and strategies of actors to other involved actors.

When the collaborative process has started, strategies for *exploring content* are important to clarify the goals and perceptions of actors (Fischer, 2003) but also to build (packages) of goals and creative solutions that: (a) keep the actors interested in the process, and (b) are able to build coalitions of support among involved actors (McGuire and Agranoff, 2011; Koppenjan and Klijn, 2016). It is important how knowledge and information are used, and especially how the discussion about creative solutions that match actors' interests is managed. Network managers thus have an information processing role, as they constantly select, transmit, and interpret relevant information originating in the organization's environment (Huxham and Vangen, 2005; Van Meerkerk and Edelenbos, 2014).

In addition, the collaborative process must also be *arranged* and guided by organizational arrangements and process rules. The managerial strategy-arranging means setting (temporary) structures for consultation, interaction, and deliberation, like project organization, communication lines, etc. (Rogers and Whetten, 1982). The transaction costs of these arrangements must be kept as low as possible (Williamson, 1996), but at the same time, the arrangements have to be acceptable to the actors involved (Klijn and Koppenjan, 2016). Various arranging forms have been described and the most well known is the distinction between shared governance, lead organization-network administrative organization (Provan and Kenis, 2008).

Another important strategy mentioned in the literature is *process agreements* that draft temporary sets of rules for interaction that structure the interactions and protect each actor's core values (Klijn and Koppenjan, 2016). The rules can be seen as ground rules for behaviour and interaction in the network that the actors in the network have (explicitly) agreed on.

From empirical research, we know that connecting and exploring network strategies seem to be the most important (Klijn et al., 2010). So, in the research, we especially looked at these two types of strategies.

Combining Contract Management and Network Management in Public-Private Collaborations

Recent research into PPPs indicates a combined effect of conditions related to contract management and network management on the performance (Warsen et al., 2019) and innovation (Callens et al., 2022) of these partnerships. Callens et al. (2022) show that in PPPs, the intrinsically unpredictable innovation process benefits from network management conditions such as exploring and connecting strategies, but contract management conditions have the potential to reinforce these network management conditions by providing room in the contract to deliberate, explore, and experiment. For instance, instead of working with a detailed project design, the contract might allow more freedom to incorporate emerging insights on how to proceed with the project. This then reinforces the impact of network management on the innovation process.

Something similar might occur in public-private collaborations. The public-private collaborations that are studied in this chapter differ from PPPs as they have a less formal contract/tender stage, are often short-term collaborations, are established in very complex service environments such as the healthcare sector, are primarily focused on producing innovative services, and are highly dependent on experiences of specialized users. Hence, these partnerships have a lot in common with public-private innovation partnerships (Di Meglio, 2013; Brogaard, 2021). However, the core mechanism of the combined effect of contract management and network management should not solely be related to PPPs. Other collaborations also try to control risk and uncertainty by establishing formal or informal agreements when the project starts, which define the expectations of the partners and the boundaries of the project, and by applying network management strategies to control complexities throughout the lifespan of the project. Hence, we propose the following hypothesis:

H1: eHealth partnerships which combine conditions of contract management, such as output specifications and contractual incentives, and conditions of network management, such as exploring and connecting strategies, generate highly innovative eHealth services.

CASES AND METHODOLOGIES

Case Selection

The European Union emphasizes the importance of digital solutions in the health sector and pushes its member states to adopt new eHealth technologies (European Commission, 2018). However, knowledge about collaborative innovation in these eHealth technologies is still limited (Wass and Vimarlund, 2016). For this reason, we selected eHealth partnerships in this study. A total of 19 eHealth partnerships from five different European countries (Belgium, the Netherlands, Denmark, Spain, and Estonia) were selected. The same case sample was used for Chapter 8, to which we refer for an elaboration on the case selection criteria. A detailed overview of the selected cases can be found in the Appendix (Table 9A.1), or on the TROPICO[1] case study repository (https://tropico-project.eu/case-studies/).

Fuzzy-Set Qualitative Comparative Analysis

The case data is analyzed through fuzzy-set qualitative comparative analysis (fsQCA), which is a case-sensitive method that uses Boolean logic to infer patterns between certain conditions and an outcome (Ragin, 2008). We used this methodology because it allows us to examine the combined effect of conditions on a certain outcome (here 'innovativeness'). The method also enables a qualitative comparative analysis between more cases than is feasible with in-depth qualitative analyses (i.e., medium N-sized samples), which is required if we want to gain insights on public-private eHealth partnerships in Europe. The downside of this method is that it uses its own terminology (e.g., 'conditions' and 'outcome' instead of respectively 'independent variables' and 'dependent variable') and is sometimes quite technical. For the purpose of this chapter, we only explain the features of fsQCA that are necessary to understand the results. A more detailed introduction to the methodology can be found in the handbook of Schneider and Wagemann (2012).

Through fsQCA, researchers can uncover if conditions (thus in our case output specifications, contractual incentives, exploring strategies, and connecting strategies) are necessary or sufficient for a particular outcome (in our case innovativeness). A condition is *necessary* when the outcome is always present when the condition is present. A (combination of) condition(s) is *sufficient* when it consistently leads to the outcome. By determining the overlap between sets of conditions and the set of the outcome, one can uncover the necessity and sufficiency of the conditions for this outcome.[2] The fsQCA analyses (i.e., analysis of necessity and analysis of sufficiency) subsequently calculate how much the sets overlap with the set of the outcome, which shows how strong the relationship between the (combination of) condition(s) and the outcome is. For this, two measures are determined, that is, *consistency*, which calculates the degree of overlap between the sets, and *coverage*, which determines the number of cases that are covered by this overlap.

Data Collection

Data was collected through a collaborative endeavour in the five selected countries. Each research team gathered data for their country, which culminated in a dataset of more than 130 observations. More specifically, 132 semi-structured interviews were conducted with project coordinators, public actors (e.g., governments, public hospitals, etc.), private actors (e.g., firms, non-profits, consultants, etc.), and service users (e.g., physicians, patients, medical professionals, etc.). Prior to the interviews, survey data was collected from 124 of these respondents through an online questionnaire. The use of multiple data collection instruments (and types of respondents) reduces the likelihood of common method/source bias but also has direct advantages for the calibration procedure and analysis of the result. On the one hand, the survey data allowed a standardized data gathering, which enabled a systematic and consistent calibration of the data. On the other hand, the interview data enriched the calibration procedure with in-depth qualitative data and made it possible to search for causal mechanisms of the discovered fsQCA patterns.

However, the use of multiple data sources also made the calibration procedure more challenging, as proper data triangulation is necessary. To remain highly consistent in our calibration, we chose to collect the interview data in a standardized manner. Each research team filled out an extensive questionnaire in which they added all the necessary data from the interviews per condition/item. To capture the remaining contextual information that was neglected due to the standardized approach, each research team also wrote a concise summary for each case. As such, all the necessary data could be collected in a semi-standardized manner and centralized for the purpose of calibration.

Measurement of the Outcome and Conditions

Innovation is defined in this chapter as 'an idea, practice or object that is perceived as new by an individual or other unit of adoption' (Rogers, 2003, p. 12). The concept is therefore composed of two elements. On the one hand, the perceived newness of innovation is an important element to distinguish something innovative from something that is not innovative (de Vries et al., 2015). On the other hand, innovation is, in contrast to related concepts such as creativity and invention, something that is implemented in a real-life environment (Walker, 2007; Anderson et al., 2014). For this reason, we measured both the degree of novelty and the level of adoption, using a bipolar, seven-point scale. The specific items are illustrated in the Apppendix (Table 9A.3). Some of these items were asked in the survey, while others were asked during the interviews. Factor analysis showed that all the items loaded on the same factor, which urged us to calculate the mean value of these items for each respondent.

The conditions were measured in similar ways. The contractual conditions 'output specifications' and 'contractual incentives' were measured on a seven-point scale (1 – Not at all; 7 – Completely), respectively, with the questions: 'The written agreement described in a detailed way the features of the innovation that had to be developed' and 'The stipulations in the written agreement or procurement criteria provided strong stimuli for developing something innovative'. The network management conditions 'exploring' and 'connecting' were measured through validated items of Klijn et al. (2010), measured through a bipolar, seven-point scale. An example of the used items for 'exploring' is 'There has been a lot of attention for involving external organizations who could bring in new ideas', and an example for 'connecting' is 'In case of deadlocks and problems in the project, it was tried as much as possible to align opposing interests'. The entire constructs for these conditions are shown in the Appendix (Table 9.A.4).

As mentioned earlier, each case is assigned to sets for the conditions and a set for the outcome. High set membership means that the case has high levels of a specific condition or outcome. To ensure a coherent calibration, the calibration procedure was centralized to one research team. The calibrated dataset is shown in the Appendix (Table 9A.2). A detailed overview of the calibration steps for each condition/outcome is highlighted in the Appendix (Table 9A.8).

Results

We made use of the fsQCA software package version 3.1b to perform the analysis (Ragin, 2017). The results are reported using standards of practice (Schneider and Wagemann, 2012). First, the analysis of necessary conditions is discussed, after which the analysis of sufficient conditions is addressed. As we are particularly interested in the combined effect of contractual conditions and network management conditions, the analysis of necessary conditions is only shortly discussed. Before we do this, however, it is useful to first examine the distribution of the cases in the set of the outcome. Figure 9.1 shows the distributions of the cases in the set of 'high innovativeness' of the generated eHealth technologies. As is visible from the figure, there is a relatively even distribution between the countries, the types of healthcare systems, and the type of innovation. Regarding the latter, seven of the highly innovative eHealth services were aimed at innovating the information flows between patients, professionals, and government (e.g., integrated data sharing platforms, central communication, and monitoring systems), while five were aimed at innovating the end product/service itself (e.g., technologies based on motion sensors, mobile apps, smart cameras, and robotics).

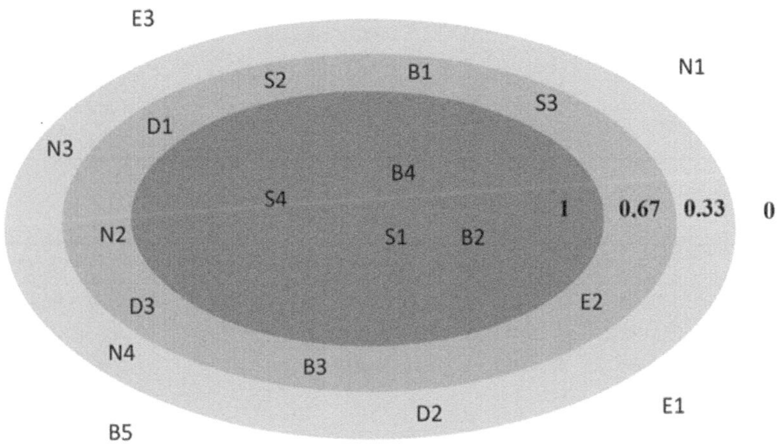

Figure 9.1 *Distribution of cases in the set of the outcome 'highly innovative eHealth services'*

We first performed the analysis of necessary conditions, both for the presence and absence (~) of the conditions. In fsQCA, a condition is regarded as 'necessary' if the condition is always present when the outcome is present A consistency threshold of 0.90 is advised when assessing the necessity of conditions (Schneider and Wagemann, 2012). As is clear from Table 9.2, none of the conditions reaches a consistency value of 0.90, and, thus, none of the conditions can be considered to be necessary for the creation of highly innovative services. Similar results arise for the absence of the outcome (Appendix, Table 9A.4).

Second, the analysis of sufficiency is performed. The first step in this analysis is the construction of a truth table, which lists all the logically possible combinations of conditions (Ragin, 2008). The truth table is shown in the Appendix (Table 9A.6). The first three rows of the truth table were retained for further analysis.[3] The second step in the analysis of sufficiency relates to the calculation of the final results.[4] Table 9.3 reports the results of this analysis. The results show that partnerships that employ output specifications and

Table 9.2 Analysis of necessary conditions

Presence of highly innovative services		
Conditions	**Consistency**	**Coverage**
Output specifications	0.664669	0.769053
~Output specifications	0.499999	0.484526
Contractual incentives	0.665667	0.645692
~Contractual incentives	0.499001	0.576701
Exploring	0.866266	0.721529
~Exploring	0.496007	0.713056
Connecting	0.731536	0.628106
~Connecting	0.531935	0.727149

Table 9.3 Results for the presence of highly innovative services

	Consistency	**Raw coverage**	**Unique coverage**	**Cases in path**
Output specifications * Exploring * ~Connecting	1	0.364271	0.033932	S3, S4
Output specifications * Contractual incentives * Exploring	0.941593	0.530937	0.200598	B4, S3, B1, B2
Solution consistency	**0.944908**			
Solution coverage	0.564869			

exploring strategies but do not use connecting strategies (~) generate highly innovative eHealth services. In addition, the results also indicate that partnerships that use output specifications, contractual incentives, and exploring strategies also generate highly innovative eHealth services. With a very high solution consistency of *0.94*, the empirical data shows that these combinations of conditions consistently lead to the outcome.[5]

Note that these results relate to the intermediate solution (QCA has three types of solutions, that is, the parsimonious solution, the complex solution, and the intermediate solution (Schneider and Wagemann, 2012), which considers the directional expectations that relate to our theoretical assumption in H1.

According to Schneider and Wagemann (2012), potential causal relations between the conditions and the outcome should always be interpreted using qualitative case information. The case information of the two covered cases in the first combination of conditions shows that the presence of a contract was important to align the differences in objectives and perspectives, but that stimulating the exploration of each other's ideas and knowledge was also indispensable. Indeed, the contract did not stimulate the emergence of collaborative synergies, which could be used to create new, innovative ideas, which was the reason for the presence of the exploring strategy. As the contract, and more specifically, the output specifications, had a large impact on the alignment of objectives and perspectives, the connecting strategy was not needed in these partnerships.

We see something similar in the second combination of conditions, in which the connecting strategy could be present or absent in the covered cases, and which means that there is no clear relationship between this condition and the outcome (at least not in this configuration of conditions). In these cases, we also see a strong presence of contract management, as not only output specifications but also contractual innovation incentives are present in these cases. In all the covered cases, contract management provided the framework of participation, which made an explicate use of connecting strategies not always necessary. However, contract conditions could not incentivize the partners enough to produce innovation during the innovation process; they needed exploring strategies during the innovation process to come up with innovative ideas.

DISCUSSION

Our findings indicate that our hypothesis, that is, that combinations of contract management, like output specifications and contractual incentives, and network management, such as exploring and connecting strategies generate highly innovative eHealth services, is only partially confirmed. Indeed, although we find that both conditions related to network management and conditions related to contract management are combined, different combinations of these conditions are present. Particularly the combination of output specifications and exploring stand out in our results, as this combination is present in both solution paths (Table 9.3). These findings are partly a confirmation of earlier research findings, both in PPP and in governance research, but also deviate from those, in the sense that particular combinations of conditions seem to exist in these innovation-oriented partnerships. Below, we reflect on related literature and provide some final conclusions.

Findings Related to Earlier Research

That we need a combination of more 'soft' managerial conditions combined with 'harder' (contractual) conditions to get good outcomes in PPPs was already known. For instance, Warsen et al. (2019) and Callens et al. (2022) showed this using a larger number of cases and QCA, both for performance and innovation. The importance of network management strategies has been shown in survey research (Klijn et al., 2010; Cristofoli et al., 2019). In that sense, our findings contribute to the ongoing empirical data we have received over the last decade about the performance and innovation of PPPs. The data consistently shows that the original idea of PPPs at the start, to achieve good performance, get the contract right, and monitor and punish (Savas, 2000; Hodge and Greve, 2010) is not the correct answer. This is related to the complexity of public-private collaborations. Public and private partners engage in relations that last for a longer period, and the ongoing interactions in the project also contain unexpected events that cannot be met and be foreseen by contract rules. The output specifications can never be enough to achieve good performance or innovation since they are unable to cover unforeseen events, difficulties emerging in the project, and new ideas that come up during the project. Given this character of public-private collaborations, it is logical that we find solution paths that both contain managerial activities and contract characteristics. This insight is very important because it shows that these collaborations have to be managed despite their contractual arrangements, and thus they need nurturing and effort to function.

In previous research, connecting as a management strategy is found to be important (Klijn et al., 2010; Warsen et al., 2019; Callens et al., 2022), while in this research it was not. This may very well be related to the specificities of the innovation process in the studied eHealth partnerships. Innovation thrives in turbulent and creative environments, in which connecting strategies might lead to premature closure of the idea-generation phase, which is detrimental to innovation (Basadur et al., 2012; Sousa et al., 2014). The exploring strategy, however, will stimulate idea divergence, which triggers idea generation and innovation (Puccio and Cabra, 2012; Burch et al., 2019). However, it may also be that there is another explanation for our finding. In comparison to previous research on PPPs, in which the connecting strategy also seems to be required to generate innovation (Callens et al., 2022), the studied eHealth partnerships had a shorter lifespan, were smaller and more focused on specific eHealth services, and emphasized experimentation, trial-and-error, and creative ideation. Especially the smaller size of these partnerships compared to those earlier studied may explain the lack of importance of the connecting strategy. After all, intensive connecting strategies are not necessary in smaller partnerships where contacts are usually more frequent and fewer actors need to be connected anyhow.

FINAL CONCLUSIONS

Our research seems to indicate that other conditions are more important for innovation than for performance, and other conditions are more important in public-private eHealth partnerships than in, for instance, infrastructural PPPs, which is the research topic of the majority of the available public-private collaboration research. Other Dutch research, for instance, showed that larger

PPP projects usually have more innovation but perform less well than smaller projects (Koppenjan et al., 2022), and that network management is less effective for innovation than for performance (Klijn et al., 2023). Future research might look beyond the well-known assumptions of both economic theory and governance theory and explore other conditions than contract characteristics and management to fully unravel the puzzle of innovations. One can think of conditions that have to do with the characteristics of the partners (are they similar or different), characteristics of the innovation process (i.e., idea generation versus idea implementation), or characteristics that are connected to the nature of the innovation (e.g., technological sophistication of ICT-enabled service innovations). Such research might further unravel the core dynamics of cross-sectoral collaboration for innovation.

NOTES

1. The TROPICO project received funding from the European Union's Horizon 2020 research and innovation programme under grant agreement No. 726840. For more information: https://cordis.europa.eu/project/id/726840.
2. Using the case data, cases are assigned to the sets of conditions and the outcome. During the calibration procedure, these set memberships are determined. Four types of set membership are used in this chapter, that is, '0' for full non-membership, '1' for full membership, '0.33' for partial non-membership, and '0.67' for partial membership. For instance, a case in which the empirical data shows that a certain condition is absent receives a membership score of 0. This calibration is performed for each condition and for the outcome (thus a case that shows no innovation is scored with 0).
3. We only retained the three first rows for the subsequent step in the analysis because of several reasons (see also Schneider and Wagemann, 2012). First, a raw consistency value of at least 0.80 is advised to select truth table rows, which excludes rows 6–11. Second, a substantial drop in consistency is observed between row 3 and row 4, which indicates that the consistency threshold is reached. Third, a contradictory case (i.e., a case that is present in the set of the solution path, but not in the set of the outcome) is present in row 4, which indicates that the empirical information is not entirely solid. Fourth, the PRI consistency (Proportional Reduction in Inconsistency), which is a measure of the degree to which the row not only leads to the presence of the outcome but also to the absence of the outcome, drops substantially in row 4 (Mendel and Ragin, 2011). Fifth, the product of the raw consistency and the PRI consistency for row 4 (0.50) is much lower than for rows 1–3 (respectively 1; 1; 0.77) (Mendel and Ragin, 2011), which indicates that only rows 1–3 consistently lead to the presence of the outcome.
4. The calculation of the final results is performed through the logical minimization of the truth table rows and the execution of the Standard Analysis, which estimates, based on the three remaining truth table rows, which combination(s) of conditions is/are sufficient for the outcome (Schneider and Wagemann, 2012).

5. There are also no contradictory cases in these combinations of conditions (i.e., cases that are covered by the combination of conditions but do not exhibit the outcome), and there is no model ambiguity (i.e., multiple tied prime implicants).

REFERENCES

Agranoff, R., and McGuire, M. (2003). *Collaborative Public Management: New Strategies for Local Governments.* Washington, DC: Georgetown University Press.

Akintoye, A., Beck, M., and Hardcastle, C. (2008). *Public-Private Partnerships – Managing Risks and Opportunities*. Oxford: Blackwell Science.

Anderson, N., De Dreu, C.K.W., and Nijstad, B.A. (2004). The routinization of innovation research: A constructively critical review of the state-of-the-science. *Journal of Organizational Behavior, 25*(2), 147–73.

Anderson, N., Potočnik, K., and Zhou, J. (2014). Innovation and creativity in organizations: A state-of-the-science review, prospective commentary, and guiding framework. *Journal of Management, 40*(5), 1297–333.

Ansell, C., and Gash, A. (2008). Collaborative governance in theory and practice. *Journal of Public Administration Research and Theory, 18*(4), 543–71.

Basadur, M., Basadur, T., and Licina, G. (2012). Organizational development. In M.D. Mumford (ed.), *Handbook of Organizational Creativity* (pp. 667–703). London: Academic Press.

Brogaard, L. (2021). Innovative outcomes in public-private innovation partnerships: A systematic review of empirical evidence and current challenges. *Public Management Review, 23*(1), 135–57.

Brown, T.L., Potoski, M., and Van Slyke, D. (2016). Managing complex contracts: A theoretical approach. *Journal of Public Administration Research and Theory, 26*(2), 294–308.

Burch, G.F., Jana, J.B., and Batchelor, J.H. (2019). Group Creative problem solving: The role of creative personality, process and creative ability. *Quality Innovation Prosperity*. doi:10.12776/QIP.V23I3.1286.

Callens, C., Verhoest, K., and Boon, J. (2022). Combined effects of procurement and collaboration on innovation in public-private-partnerships: A qualitative comparative analysis of 24 infrastructure projects. *Public Management Review, 24*(6), 860–81 https://doi.org/10.1080/14719037.2020.1867228.

Cristofoli, D., Trivellato, B., and Verzillo, S. (2019). Network management as a contingent activity: A configurational analysis of managerial behaviors in different network settings. *Public Management Review, 21*(12), 1775–800.

Damanpour, F., and Schneider, M. (2008). Characteristics of Innovation and innovation adoption in public organizations: Assessing the role of managers. *Journal of Public Administration Research and Theory, 19*(3), 495–522.

De Palma, A., Leruth, L.E., and Prunier, G. (2009). *Towards a Principal-Agent Based Typology of Risks in Public-Private Partnerships*. International Monetary Fund. https://www.imf.org/en/Publications/WP/Issues/2016/12/31/Towards-a-Principal-Agent-Based-Typology-of-Risks-in-Public-Private-Partnerships-23190.

De Vries, H., Bekkers, V., and Tummers, L. (2015). Innovation in the public sector: A systematic review and future research agenda. *Public Administration, 94*(1), 146–66.

Di Meglio, G. (2013). The place of servPPINs in the range of public-private collaboration arrangements for services provision. In Allouj, Rubalcaba, and Windrum (eds.), *Public-Private Innovation Networks in Services*, 59–87. Cheltenham, UK and Northampton, MA, USA: Edward Elgar.

Edquist, C., Vonortas, N., and Zabala-Iturriagagoitia, J.M. (2015). Introduction. In C. Edquist, N.C. Vonortas, J.M. Zabala-Iturriagagoitia, and J. Edler (eds.), *Public Procurement for Innovation* (pp. 1–34). Cheltanham, UK and Northampton, MA, USA: Edward Elgar.

Emerson, K., and Nabatchi, T. (2015). *Collaborative Governance Regimes*. Washington DC: Georgetown University Press.

European Commission (2018). *Communication from the commission to the European Parliament, the Council, the European Economic and Social Committee and the Committee of the Regions on enabling the digital transformation of health and care in the Digital Single Market; empowering citizens and building a healthier society* (COM/2018/233 final). https://eur-lex.europa.eu/legal-content/EN/TXT/?uri=COM%3A2018%3A233%3AFIN.

Fischer, F. (2003) *Reframing Public Policy: Discursive Politics and Deliberative Practices*. Oxford: Oxford University Press.

Flyvbjerg, B., Holm, M.S., and Buhl, S. (2003). How common and how large are cost overruns in transport infrastructure projects? *Transport Reviews, 23*(1), 71–88.

Gage, R.W., and Mandell, M.P. (Eds.) (1990). *Strategies for Managing Intergovernmental Policies and Networks*. New York: Praeger.

Geroski, P. (1990). Procurement policy as a tool of industrial policy. *International Review of Applied Economics, 4*(2), 182–98.

Georghiou, L., Edler, J., Uyarra, E., and Yeow, J. (2014). Policy instruments for public procurement of innovation: Choice, design and assessment. *Technological Forecasting and Social Change, 86*, 1–12.

Grimsey, D., and Lewis, M. (2007). Public private partnerships and public procurement. *Agenda: A Journal of Policy Analysis and Reform, 14*(2), 171.

Hale, J.S., and Woronkowicz, J. (2021). Artists as public sector intrapreneurs: An experiment. *Small Business Economics, 57*, 653–66.

Hodge, G.A., and Greve, C. (2010). Public-private partnerships: Governance scheme or language game? *Australian Journal of Public Administration, 69*(S1), S8–S22.

Huxham, C., and Vangen, S. (2005). *Managing to Collaborate: The Theory and Practice of Collaborative Advantage*. Abingdon, UK: Routledge.

Jensen, M.C., and Meckling, W.H. (1976). Theory of the firm: Managerial behavior, agency costs and ownership structure. *Journal of Financial Economics, 3*(4), 305–60.

Kane, G.C. (2015). How digital transformation is making health care safer, faster and cheaper. *MIT Sloan Management Review, 57*(1), 1–11.

Kickert, W.J.M., Klijn, E.H., and Koppenjan, J.F.M. (1997). *Managing Complex Networks: Strategies for the Public Sector*. London: Sage.

Klijn, E.H., and Koppenjan, J. (2016). The impact of contract characteristics on the performance of public–private partnerships (PPPs). *Public Money and Management, 36*(6), 455–62.

Klijn, E.H., Steijn, B.. and Edelenbos, J. (2010). The impact of network management strategies on the outcomes in governance networks. *Public Administration, 88*(4), 1063–82.

Klijn, E.H., Metselaar,S., and Warsen, R. (2023). The effect of contract and network management on performance and innovation in infrastructure projects. *Public Money and Management*. doi:10.1080/09540962.2023.2204533.

Koppenjan, J.F.M, Klijn, E.H., Verwij, S., Duijn, R., van Meerkerk, M., Metselaar, I.S, and Warsen, R. (2022). The performance of public-private partnerships: An evaluation of 15 years DBFM in Dutch infrastructure. *Governance Public Performance and Management Review, 45*(5), 998–1028. https://doi.org/10.1080/15309576.2022.2062399.

Kraus, S., Schiavone, F., Pluzhnikova A., and Invernizzi, A.H. (2021). Digital transformation in healthcare: Analyzing the current state-of-research. *Journal of Business Research, 123*, 557–67.

Lehne, M., Sass, J., Essenwanger, A., Schepers, J., and Thun, S. (2019). Why digital medicine depends on interoperability. *Digital Medicine, 79*, 1–5.

Loveys, K., Fricchione, G., Kolappa, K., Sagar, M., and Broadbent, E. (2019). Reducing patient loneliness with artificial agents: Design insights from evolutionary neuropsychiatry. *Journal of Medical Internet Research, 21*(7). doi:10.2196/13664.

Mantel, S.J. (2005). *Core Concepts of Project Management* (2nd edn). New York: Wiley.

McGuire, M., and Agranoff, R. (2011). The limitations of public management networks. *Public Administration, 89*(2), 265–84.

Mendel, Jerry M., and Ragin, Charles C. (2011). fsQCA: Dialog between Jerry M. Mendel and Charles C. Ragin. USC-SIPI REPORT No. 411.

Mergel, I., Edelmann, N., and Haug, N. (2019). Defining digital transformation: Results from expert interviews. *Government Information Quarterly, 36*, 101385.

Nooteboom, B. (2002). *Trust: Forms, Foundations, Functions, Failures and Figures.* Cheltenham, UK and Northampton, MA, USA: Edward Elgar.

Pennisi, P., Tonacci, A., Tartarisco, G., Billeci, L., Ruta, L., Gangemi, S., and Pioggia, G. (2016). Autism and social robotics: A systematic review. *Autism Research, 9*(2), 165–83.

Puccio, G.J., and Cabra, J.F. (2012). Idea generation and idea evaluation: Cognitive skills and deliberate practices. In M.D. Mumford (ed.), *Handbook of Organizational Creativity* (pp. 667–703). London: Academic Press.

Provan, K.G., and Kenis, P. (2008). Modes of Network governance: Structure, management, and effectiveness. *Journal of Public Administration Research and Theory, 18*(2), 229–52.

Ragin, C. (2008). *Redesigning Social Inquiry: Fuzzy Sets and beyond.* Chicago, IL: University of Chicago Press.

Rogers, D.L., and Whetten, D.A. (eds.) (1982). *Inter-Organizational Coordination: Theory, Research, and Implementation.* Ames, IA: Iowa State University Press.

Rogers, E.M. (2003). *Diffusion of Innovations* (5th edn). New York: Free Press.

Savas, E.S. (2000). *Privatization and Public–Private Partnerships.* New York: Chatham House.

Scharpf, F.W. (1978). Inter-organizational policy studies: Issues, concepts and perspectives. In K. Hanf and F. Scharpf (eds.), *Interorganisational Coordinations: Limits to Coordination and Control* (pp. 345–70). London: Sage.

Schneider, C.Q., and Wagemann, C. (2012). *Set-Theoretic Methods for the Social Sciences: A Guide to Qualitative Comparative Analysis.* New York: Cambridge University Press.

Sørensen, E., and Torfing, J. (2011). Enhancing collaborative innovation in the public sector. *Administration and Society, 43*(8), 842–68.

Sousa, F.C., Monteiro, I.P., Walton, A.P., and Pissarra, J. (2014). Adapting creative problem solving to an organizational context: A study of its effectiveness with a student population. *Creativity and Innovation Management, 23*(2), 111–20.

Steinhubl, S.R., Muse, E.D. and Topol, E.J. (2015). The emerging field of mobile health. *Science Translational Medicine, 7*(283). doi: 10.1126/scitranslmed.aaa3487.

Steijn, A.J., Klijn, E.H., and Edelenbos, J. (2011). Public Private partnerships: Added value by organizational form or management? *Public Administration, 89*(4), 1235–52.

Torfing, J. (2019). Collaborative innovation in the public sector: The argument. *Public Management Review, 21*(1), 1–11.

Torfing, J., Cristofoli, D., Gloor, P.A., Meijer, A.J., and Trivellato, B. (2020). Taming the snake in paradise: Combining institutional design and leadership to enhance collaborative innovation. *Policy and Society, 39*(4), 592–616.

Uyarra, E., Edler, J., Garcia-Estevez, J., Georghiou, L., and Yeow, J. (2014). Barriers to innovation through public procurement: A supplier perspective. *Technovation, 34*(10), 631–45.

van Meerkerk, I., and Edelenbos, J. (2014). The effects of boundary spanners on trust and performance of urban governance networks: Findings from survey research on urban development projects in the Netherlands. *Policy Sciences, 47*, 3–24.

Walker, R.M. (2007). An empirical evaluation of innovation types and organizational and environmental characteristics: Towards a configuration framework. *Journal of Public Administration Research and Theory, 18*, 591–615.

Warsen, R., Klijn, E.H., and Koppenjan, J.F.M. (2019). Mix and match: How contractual and relational conditions are combined in successful public–private partnerships. *Journal of Public Administration Research and Theory, 29*(3), 375–93.

Wass, S., and Vimarlund, V. (2016). Healthcare in the age of open innovation – a literature review. *Health Information Management Journal, 45*(3), 121–33.

Williamson, O.E. (1996). *The Mechanisms of Governance*. Oxford: Oxford University Press.

APPENDIX

Table 9A.1 Selected cases

Cases	Collaborative eHealth innovation
Belgium	National portal website which provides information for all the citizens, created through a collaboration between government agencies, ministerial cabinets, hospital networks, regional governments, private health suppliers, and insurance organizations, and user organizations
Mixed Napoleonic – Etatist Social Health Ins.	Patient information sharing tool for GPs and home care organizations, created through a collaboration between private nursing organizations and federation, ministerial cabinets, national government agencies, hospital networks, individual GPs, and several private health organizations
	A way of creating, validating, and disseminating official evidence-based guidelines for healthcare providers, created through a collaboration between universities, private health organizations, national and regional government agencies, red cross organizations, knowledge organizations, ICT suppliers, and individual health professionals
	Several technologies in a nursing home (wearables, smart cameras, etc.), created through a collaboration between a public nursing home (local government), private construction companies and contractors, consultant companies, nurses, and patients
	Platform which brings people with health/social care demands together with volunteers, created through municipalities, communal network, private hospitals, private ICT companies, consultant companies, citizens, and health professionals
The Netherlands	ICT platform which facilitates the exchange of health information between partners and patients, created through the collaboration between a municipality, public hospital, and several private health organizations
Continental adm. Regime	Digital platform designed to foster neighbourhood collaborations between clients and consultants, created through the collaboration between a municipality, private healthcare provider, neighbourhood teams, citizens
Etatist Social Health Ins.	Tracking technologies in a nursing home, created through the collaboration between a semi-private association, software developer, and patient organization
	'Smart diaper' for elderly people, created through the collaboration between a semi-private association, ICT company, consultant company

Cases	Collaborative eHealth innovation
Spain	Electronic prescription system, patient appointment system, robot for automatic storage and dispensing, created through the collaboration between several public hospitals, private ICT companies, several patient organizations, university
Napoleonic – National Health Service	Home health ICT tools for chronic patients, created through the collaboration between a public hospital/health service, regional government, ICT companies, consultancy companies, several other private companies, universities, health professionals and patients
	Web application for computerized cognitive behaviour therapy (CCBT), created through the collaboration between public hospitals and healthcare services, public research institute, private technology centre, several health professionals (e.g., psychiatrist, psychologists, physicians, etc.)
	AI used to diagnose uncooperative patients, created through the collaboration between public hospitals, ICT and telecom companies, physicians
Estonia	Centralized patient registration system, created through a collaboration between the ministry, government agencies and public authorities, ICT companies, private healthcare providers, physician associations, hospital associations, individual physicians
Eastern European – Etatist Social Health Ins.	Integration of application processes for rehabilitation, disabilities, aids, created through a collaboration between ministries, public health insurance authority, government agencies, physician association, interest groups
	Voice command app to guide healthcare providers, created through a collaboration between a ministry, public health insurance authority, colleges, network of healthcare providers, ICT companies, several healthcare organizations
Denmark	E-learning programme regarding dysphagia, created through a collaboration between a regional government, municipalities, public hospitals, ICT company, representatives of health professionals
Nordic – National Health Service	Smartphone app for patient reported outcomes, created through a collaboration between a public hospital, ICT company, health professionals
	Smartphone app that helps convey the results of bone scans to patients with osteoporosis, created through a collaboration between a public hospital, university, ICT and health service companies, patient associations, health professionals

Table 9A.2 Calibrated dataset

Case	Output specifications	Contractual incentives	Exploring	Connecting	Perceived innovativeness
N3	1	0.67	0.33	0.67	0.33
B5	0	1	0.67	0.33	0
E1	0.67	0.33	0.67	1	0
E3	0	0.33	0.67	0.67	0
D1	0.33	1	0.67	1	0.67
B3	0.33	0.33	0.67	0.33	0.67
N4	0	0	0.67	1	0.33
N2	0	0	0.33	0.33	0.67
S3	0.67	0.67	0.67	0.33	0.67
B1	1	1	0.67	1	0.67
B2	1	1	0.67	0.67	0.67
D3	0.33	0.33	0.67	0.33	0.67
S2	0.33	0.67	1	0.67	0.67
E2	0	0	0.67	0.67	0.67
D2	0	0	0.33	1	0.33
S1	1	1	0.33	0.67	1
S4	1	0.33	0.67	0.33	1
B4	1	1	1	0.67	1
N1	0	0.67	0.67	0	0

Table 9A.3 Operationalization of innovativeness

Newness	Adoption
No/A lot of innovative ideas are developed in this project	The frequency of use will typically be very low/high
The innovativeness of the developed innovation is very low/high	The effect on a user's life will be very small/extensive
The innovative character of the project is lower than/exceeds my initial expectations	Only a selective subgroup of users/All users that would benefit from this innovation can use it
The users could do exactly the same thing with other tools/would be unable to do those things without this innovation	The innovative ideas that are developed in the project are not feasible at all/very feasible
It is very easy/difficult (or impossible) to find tools that have the same functionalities as this innovation (at the moment of implementation)	The innovation does not deal with the problems at hand at all/really deals with the problems at hand

Table 9A.4 Operationalization of network management

Survey items	Interview items
Exploring strategy	
There has been a lot of attention for involving external organizations who could bring in new ideas	Did actors in the project try to reveal as much as possible different perspectives and integrate them into the decision-making? YES/NO + examples
When gathering information and knowledge in this project, there was a lot of emphasis on determining the joint information needs	Did actors in the project stimulate the exploration/ acquisition of new information/knowledge or the inclusion of new actors with such information or knowledge? YES/NO + examples
It has been attempted to include as much as possible different opinions and perspectives in the decision-making	
Connecting strategy	
In case of deadlocks and problems in the project, it was tried as much as possible to align opposing interest	Was it tried in the project to align/reconcile opposing views or opinions? YES/NO + examples
Besides the coordinator(s), all the other actors were involved actively in taking decisions	Confronted with conflicts or deadlocks, was it attempted to enhance the interactions between opposing actors or to weaken the interactions between opposing actors. Enhance/Weaken + examples
In this project there has been a lot of attention for the relationships between the involved individuals and organizations	

Table 9A.5 Analysis of necessary conditions – absence of highly innovative services

Absence of highly innovative services		
Conditions	**Consistency**	**Coverage**
Output specifications	0.406460	0.421478
~Output specifications	0.777285	0.675048
Contractual incentives	0.591315	0.514037
~Contractual incentives	0.592429	0.613610
Exploring	0.777285	0.580216
~Exploring	0.626950	0.807747
Connecting	0.777285	0.598115
~Connecting	0.516705	0.633015

Table 9A.6 Truth table

	Output specifications	Contractual incentives	Exploring	Connecting	Innovation[a]	#cases	Raw consist.	PRI consist.
1	1	0	1	0	**1**	1	**1**	**1**
2	1	1	1	0	**1**	1	**1**	**1**
3	1	1	1	1	**1**	3	**0.933735**	**0.858974**
4	1	1	0	1	0	2	0.831658	0.598802
5	0	0	1	0	0	2	0.801205	0.507463
6	0	0	0	0	0	1	0.750943	0.34
7	1	0	1	1	0	1	0.74717	0.33
8	0	1	1	1	0	2	0.701807	0.407186
9	0	0	0	1	0	1	0.697885	0
10	0	0	1	1	0	3	0.689815	0.202381
11	0	1	1	0	0	2	0.441472	0

Note: [a] The 1 in the columns indicates that rows 1 through 3 consistently lead to the outcome.

Table 9A.7 Parsimonious solution for the presence of highly innovative services

	Consistency	Raw coverage	Unique coverage	Cases in path
Output specifications * Exploring * ~Connecting	1	0.364271	0.033932	S3, S4
Output specifications * Contractual incentives * Exploring	0.941593	0.530937	0.200598	B4, S3, B1, B2
Solution consistency	**0.944908**			
Solution coverage	0.564869			

Table 9A.8 Calibration of outcome/conditions

Innovation of services Output specifications Contractual incentives	**Exploring Connecting**
Survey data leading	***STEP 1: Survey data***
All answers of the respondents above the cross-over point → 1 More than half of the answers above the cross-over point → 0.67 More than half of the answers below or on the cross-over point → 0.33 More than half of the answers below the cross-over point → 0 Equal amount above and below/on the cross-over point → Larger distance to the cross-over point of answer resp. above and below/on cross-over point is indicative for assigning case score above or below cross-over point (i.e., 0/0.33 or 0.67) + qualitative interpretation to assign 0 or 0.33	All answers of the respondents above the cross-over point → 1 More than half of the answers above the cross-over point → 0.67 More than half of the answers below or on the cross-over point → 0.33 More than half of the answers below the cross-over point → 0 Equal amount above and below/on the cross-over point → Larger distance to the cross-over point of answer resp. above and below/on cross-over point is indicative for assigning case score above or below cross-over point (i.e., 0/0.33 or 0.67) + qualitative interpretation to assign 0 or 0.33
General qualitative check of the assigned scores using the interview data	
	STEP 2: Interview data
	Number of examples of the network management strategy from the interviews
	0 or 1 examples → 0 2 or 3 examples → 0.33 4 or 5 examples → 0.67 6 or more examples → 1
	STEP 3: Qualitative interpretation
	Using qualitative case information extracted from the interviews, a third score is assigned by the researcher

Innovation of services	Output specifications	Contractual incentives	Exploring	Connecting
			STEP 4: Final score	
			Calculate mean of survey and interview score If not exactly 0;0.33;0.67;1, round towards the qualitative interpretation If qualitative interpretation is lower than cross-over point and mean is higher, case score will always be lower than cross-over point (e.g., mean = 0.67, no rounding necessary, but qualitative interpretation is 0, case score becomes 0.33). Idem when qualitative interpretation is higher than cross-over point and mean is lower	

10. Viewpoints of collaboration partners on user involvement in collaborative innovation projects

Jaime García-Rayado and Chesney Callens

INTRODUCTION

In the last decades, the idea of an interdependent network of public and private institutions, citizens, and third sector organizations participating in policy preparation processes and public service delivery has been gaining support under the paradigm of the New Public Governance (NPG) (Klijn and Koppenjan, 2012; Osborne, 2010; Pestoff et al., 2012). Collaboration has been seen as a solution for the improvement of public services and the recovery of legitimacy. Indeed, collaborative strategies can enhance innovation better than traditional hierarchical and competitive strategies (Torfing, 2019). Collaboration can reinforce all parts of the innovation process, as it allows organizations to access knowledge, resources, experiences, and perspectives that are different from their own (Sørensen and Torfing, 2011). However, collaboration also implies new challenges such as the design of mechanisms for active, balanced, and continuous multi-stakeholder participation or the achievement of efficient coordination between actors with different objectives (Koppenjan, and Koliba, 2013; Osborne, 2010; Torfing and Trianfillou, 2013).

In a collaborative innovation project, two types of actors are typically engaged in the partnership. On the one hand, the partners are the actors who are centrally engaged in the project, and responsible for creating a new public service. These actors can be public organizations or private companies, but also other non-state organizations or individual actors. On the other hand, the service users are the actors who are going to use the newly created service or are currently using similar services. The users can be directly involved in the partnership (e.g., as part of a project team, in testing phases, etc.) or indirectly (e.g., through their adoption of the innovation), but they are primarily characterized as the real-life users of the created service.

These differences in partners' contexts and related levels of involvement of users enrich the partnership, but can also cause friction during the interaction process, particularly in the case where the users are directly engaged in the partnership. For instance, partners and users might disagree in tasks, processes, and roles, or value-related conflicts may arise (Mele, 2011). In the first phases of a project, role-related conflicts can appear as there are differences in the involvement and responsibility that each actor thinks they should have (Mele, 2011). Indeed, each public administration paradigm reveals a different perspective on the users' role in public service delivery (Torfing et al., 2020). Because of these different frames, collaboration partners may have different opinions than users on user involvement, and a conflict that blocks the project may appear.

The purpose of this chapter is to analyse the perspectives of partners of collaborative innovation projects about the involvement of users in partnerships for innovation development. By applying Q-methodology, we contrast four roles of user-provider interaction in co-innovation processes with users (legitimators, customers, partners, and self-organizers) (Callens et al., 2023). The analysis focuses on partnerships for the development of eHealth innovations. These innovations require the collaboration of multiple actors due to the specific knowledge of different areas necessary for their development, the high validation requirements applied, and the complexity of the sector. Many eHealth innovations get stuck in the pilot phase and are never implemented (Andreassen et al., 2015). Previous research has found that to avoid this blockage, it is necessary that partnerships are capable of creating stakeholder networks to integrate resources, mainly knowledge, and have a leadership promoting the interaction and integration of ideas, especially in projects involving user communities (Urueña et al., 2016). Many of the barriers found in the implementation of eHealth innovations are related to user worries and problems, for example, lack of trust, insufficient technological skills, low impact in service quality, and time necessary to learn and use the innovation (Jang-Jaccard et al., 2014). These barriers show that the partnerships for the development of eHealth innovations are not properly integrating user knowledge or achieving a successful collaboration network. This chapter contributes to revealing whether this problem is caused by a low expectation of partners on the role of users or by the inconsistency in the roles expected, for example, incompatible roles or mismatching between the motives for user involvement and their role in the project.

This research was carried out within Work Package 7 of the TROPICO[1] project which studied the practices of external collaboration for service delivery and was funded by the European Union's Horizon 2020 Research and Innovation programme. The chapter is structured as follows. The second section provides the theoretical framework. The methodology applied is described in the third section. In the fourth section, the results are explained. Finally, the discussion and conclusions are presented.

THEORETICAL FRAMEWORK

Our research is based on three widely accepted public administration paradigms: Traditional Public Administration, New Public Management, and New Public Governance (Torfing et al., 2020), and the notion of self-governance, which is a wider term that comprises situations in which public and private individuals and organizations, on their own initiative, autonomously act to achieve public or collective objectives (Sørensen and Triantafillou, 2009). Each paradigm defines a role of citizens in public service delivery that delimits the possible relation between users and partners in a collaborative innovation project. We consider four roles of user involvement: users as (1) Legitimators, (2) Customers, (3) Partners, and (4) Self-organizers (Callens et al., 2023). Each of these roles supposes different motivations for the involvement, task, and responsibilities of the users, and the role of the service providers and the leader/coordinator of the partnership during the users' involvement. Whereas Callens et al. (2023) apply these roles to the viewpoints of users, this chapter reflects if these roles are also applicable to the viewpoints of the collaboration partners.

Legitimators

From the point of view of the *Traditional Public Administration* paradigm, citizens are only expected to elect public representatives and they have usually no, or very limited direct participation in public service delivery (Bryson et al., 2014; Denhardt and Denhardt, 2007). In this case, the users won't participate actively in collaborative innovation projects; they will just act as *legitimators* to protect user rights and needs and support the innovation. However, even if the involved users do not participate actively in collaborative innovation projects, they are still crucial stakeholders because they represent the common interest of the service users. The leader of the partnership (e.g., the project coordinator), therefore, should give them enough information about the innovation, the achievements of the partnership, and its inner workings, so the innovation is more likely to be accepted (Silvia, 2011). The leader should also protect the regulative framework, as the innovation is not only created for the involved users, but for all users (i.e., common interest instead of individual interest).

Customers

With the introduction of managerial approaches and the focus on efficiency, effectiveness, and competition related to *New Public Management*, citizens and public service users are viewed as individual customers (Bryson et al., 2014; Denhardt and Denhardt, 2007). In this perspective, the focus shifts to consumer satisfaction in service delivery, similar to a business approach in market competition, and the user is viewed as a *customer*. Customers look for the achievement of their individual interests and needs, and not the public interest. In this case, a partnership collaborating to develop an innovation would want to obtain users' knowledge to ensure the developed product or service complies with their individual expectations. Knowledge from customers provides unique information about users' preferences and can improve the new service outcome and the acceptance of the innovation (Mahr et al., 2014). Therefore, the partnership should consult the users about their preferences and the quality they expect from the innovation.

When the partnership involves the users as customers, the tasks of the partnership leader expand. Leaders must ensure the inclusion of input from the consumers, as organizations may fail to include them for various reasons, such as lack of resources, expertise, and time, disillusionment with previous experiences, poor organizational capacity, or organizational culture opposed to their inclusion (Olson and Bakke, 2001). Moreover, the expectations of the users define their incentives to participate (Ansell and Gash, 2008), and the coordinators should make them see the achievement of their individual goals through their collaboration in the network (Silvia, 2011), and let them see how the innovation works in practice, to ensure that users fulfil their individual interest. Indeed, compliance with the expectations of the involved stakeholders influences the success of the collaboration (Ansell and Gash, 2008).

Partners

The third potential role of the user appears when they are included in the collaborative innovation project on an equal footing with the other partners, that is, as another *partner*. This role aligns with *New Public Governance*, where co-production and co-creation of value with citizens and other non-state actors is a core feature (Brandsen and Honingh, 2016; Osborne and Strokosch, 2013; Osborne et al., 2016; Pestoff et al., 2012). This type of collaboration is necessary to achieve an efficient, effective, and democratic public sector (Pierre and Peters, 2020). Indeed, the co-creation of value can occur not only through the fulfilment of users' individual needs, but also through their participation in the design and delivery of the service and their contribution to innovation (Osborne et al., 2016).

If users are included as a partner, co-producing with the rest of the partnership, the leader should identify and incorporate those suitable members who can provide the necessary resources to the partnership (McGuire, 2006; Vangen and Huxham, 2003), in this case the lead users who are ahead of the rest in the market (Von Hippel, 1986). In addition, the partnership must deal with conflicts between partners such as possible discrepancies about their responsibility for the project and conflicts between the value systems of the users and the project organizations (Mele, 2011). Moreover, the alignment or misalignment between the partners' views and the actual role of users in a partnership could reinforce these conflicts and affect the deliberation process. Therefore, the leader should align the goals of partners and involved users and ensure the joint decision making between them for the success of the collaboration.

Self-organizers

Self-governing structures of non-governmental actors can carry out collective action without the interference of the government (Pierre and Peters, 2020). From this perspective, users act as *self-organizers*. Users can be initiators of new services and the government follows their lead (Voorberg et al., 2014) and carries out supportive actions, like providing technical knowledge and funding (Nederhand et al., 2016). Self-governing structures require a trigger that will initiate the interaction of the actors in a network (Nederhand et al., 2016). The lack of consideration of new products and services, or functionalities by organizations, serves as a trigger for users to look for innovation by themselves (von Hippel et al., 2011). These structures need trustworthy relationships, exchange of ideas, information, and experiences in open communication among the actors (Neederhand et al., 2016). The leader of the partnership uses a 'hands-off' approach to the governance of the collaboration, by limiting day-to-day interventions in the project (Sørensen and Torfing, 2009).

Comparison of User Roles

The four roles reflect different levels of user involvement in the projects according to the reasons for their inclusion, the activities they can perform, and how to effectively engage them. Users acting as *legitimators* are external to the projects without active participation in them as they are involved just to improve the acceptance of the innovation. When users' involvement is intended to improve the quality of the innovation, users should participate in the innovation process to provide the partnership with their knowledge as *customers*. In this case, in order to effectively capture and integrate users' knowledge, the partnership should increase the information about the innovation and provide the users with the opportunity to show their desires and needs. However, this role keeps users out of the design and decision-making processes of the project. In contrast, the involvement of users as *partners* implies that their knowledge is presumed to be necessary for the design of the project and that they have the capacity to make decisions at the project level, which may sometimes conflict with the wishes of potential consumers (e.g., decisions regarding the cost-quality ratio of the resulting product or service).

In these three roles, the partnership is 'above' the users, restricting or facilitating their participation. The partnership takes a step back when the users *self-organize*, and just support the users so they can develop their proposals, as the self-organizer users have the required capabilities to direct the project.

These four user roles are ideal representations of user involvement, and does not mean that they cannot be mixed (e.g., the participation of users giving their knowledge as consumers could be motivated for the necessity of the improvement of the acceptance of the innovation), or that users with different roles could not be involved in the same partnerships due to different necessities of the project (e.g., most of them participating as *legitimators* or *consumers* but a smaller group participating as *partners*) or different capacities of the users according to their background (e.g., physicians, nursing staff, or patients).

RESEARCH DESIGN

Q-methodology

This chapter makes use of Q-methodology, which was first introduced by Stephenson in 1930 as a method to distinguish features of individuals, such as their viewpoints (Burt and Stephenson, 1939). Q-methodology enables the factorization of individuals instead of variables on a population of traits, abilities, or characteristics, with the purpose of identifying differences between these individuals (Watts and Stenner, 2012). Because of its ability to differentiate between viewpoints of individuals, it has often been used to identify different discourses of respondents (van Exel and de Graaf, 2005). In recent years, the use of Q-methodology has increased because of its ability to distinguish different viewpoints of respondents regarding policy choices (e.g., Molenveld et al., 2019; Nederhand et al., 2019; Warsen et al., 2019) and citizen involvement (van Eijk and Steen, 2014; van Eijk et al., 2017).

Generally, Q-methodology is conducted in four sequential steps (Watts and Stenner, 2012). First, the *Q-set* is constructed. The Q-set is composed of statements that reflect the different discourses or viewpoints present in the population. Second, the *P-set*, or set of participants, is defined. The P-set represents the individuals that are relevant for the discourses or viewpoints addressed by the Q-set. In our case, the P-set consists of service users related to processes of collaborative service creation and innovation. Third, the statements defined in the Q-set are applied to the respondents in the P-set by conducting a *Q-sort*. During the Q-sort, the respondents rank the different statements in the Q-set according to the degree to which they reflect their own viewpoints. Fourth, the Q-sorts are analysed through Q-methodological *factor analysis* to separate the common variance between the respondents. Out of this factor analysis, different viewpoints or discourses regarding user involvement emerge.

Constructing the Q-set

The development of the Q-set needs to be rigorous and well thought through, as an inadequate collection of statements in the Q-set decreases the validity of the discourses in the P-set. Watts and Stenner (2012) propose two criteria to enhance the quality of the statements.

First, the statements must be representative of the discourses the researcher wants to cover. Scholars use different techniques to ensure this representativeness. Some researchers refer to academic literature out of which the statements are deductively constructed (e.g., Nederhand et al., 2019; Warsen et al., 2019), while other scholars use exploratory interviews on the topics of the discourses to extract the relevant statements inductively (Molenveld et al., 2019; van Eijk and Steen, 2014). The deductive method enables us to assess the theoretical roles of user involvement, as they are compared to the empirical profiles we get from the Q-sorts. This allows an evaluation of our own theoretical framework, which is the primary reason why we apply the deductive approach in this chapter. In addition, document analysis was used to check how practitioners perceive the different roles of users.

Second, gaps or overlaps between the statements might prevent each statement having an equal contribution to the Q-set. To ensure a balanced Q-set, scholars typically use multiple types of statements. Two methods are typically used. On the one hand, the typology of Dryzek and Berejikian (1993) is often used to distinguish statements according to how they are interpreted by the respondent (Molenveld et al., 2019; van Eijk and Steen, 2014). The authors distinguish between definitive statements ('concerning the meaning of terms'), designative statements ('concerning questions of facts'), evaluative statements ('concerning the worth of something that does or could exist'), and advocative statements ('concerning something that should or should not exist'). The designative and advocative statements are broadly used in public administration research (e.g., Molenveld et al., 2019; Nederhand et al., 2019; van Eijk and Steen, 2014; Warsen et al., 2019), which is why we apply these two types of statements to our Q-set. Statements can also be distinguished from each other based on the dimensions relevant to the study.

The roles of user involvement in our theoretical framework reflected three dimensions: (1) the motives of the users to be involved, (2) the activities the users conduct during their involvement, and (3) the role of the service providers during the user involvement. These considerations resulted in 24 statements, illustrated in Table 10.1. Table 10.1 presents two statements per type of dimension/role, whereby the first statement is a designative statement, while the second statement is an advocative statement.

Defining the P-set

In this chapter, the P-set is largely defined by the types of projects that were considered. We selected 19 projects from the health sector, as this is an established policy field in the co-production literature (e.g., Daya et al., 2019; Gremyr et al., 2018; Sangill et al., 2019; van Eijk and Steen, 2014). This ensured valid cases of user involvement in service creation. Furthermore, the pursuit of innovative eHealth technologies is high on the agenda of the European Commission (European Commission, 2018), while current research insufficiently explains successful eHealth innovations (Andreassen et al., 2015). An elaboration on the selection of cases can also be found in Chapter 8.

The respondents included in the P-set were representing the organizations acting as partners in the collaboration projects – both the public actors (e.g., governments, public hospitals, etc.) and private actors (e.g., non-profits, firms, etc.). A total of 50 partners were identified for the P-set.[2]

Conducting the Q-sort and Factor Analysis

To assure that the statements were valid and easy to understand they were tested in one of the countries (Belgium) through a pilot on respondents with similar profiles to the service users in our study. The sorting of the statements is usually conducted through a fixed structure. We used Q Method Software to conduct the Q-sort and used a sorting structure with values from −3 to 3, indicating the degree to which the respondents identify themselves with a particular statement. The respondents were first asked to presort the statements into three piles: statements they generally agreed with, statements they generally disagreed with, and statements towards which they were neutral. Next, the respondents were asked to rank the statements from −3 to 3. Once all responses were gathered, a factorial analysis was carried out with four criteria for the selection of the factors: Kaiser-Gutmann criterion, variance, cumulative variance, and significance of factor loadings.[3]

Table 10.1 *The Q-set*

Dimensions/Roles	Legitimator	Customer	Partner	Self-organizer
Motives to participate	Users should be involved primarily to create support for the innovation	Users want to be involved primarily to indicate what they perceive as an exquisite end product	Involved users especially want to be recognized as partners	Users should tackle user issues themselves instead of waiting for others to do it
	Users are especially involved to check whether the rights of those they represent are guaranteed	Involved users should above all check how user-oriented the innovation is	Users should be involved because they can have alternative views, useful for the other partners	Users know best how to develop and organize service delivery
Activities of involved users	The majority of users is there predominantly to listen to what the partners have to say	Involved users have to advise the partnership about how to increase user satisfaction	Users and the other partners should jointly define the problem and the solution	Users can best define problems and solutions
	Users best leave development of innovations to others	Just like a company asking its customers about its products, the partnership needs to consult the users about their preferences	Equal contributions of users and other partners (co-creation) is the only way to create relevant innovations	Users should set and guard the direction for the innovation process

Dimensions/Roles	Legitimator	Customer	Partner	Self-organizer
Role of partnership towards user involvement	The users should be well informed by the partnership because the innovation can then be easily accepted	The partnership should enable the involved users to see how the innovation works in reality	A crucial task of the partnership is to ensure joint decision making between the involved users and the other partners	The main role of the partnership is to provide the resources to develop proposals for the users
	The partnership actors are there to make sure that the input of the users and other actors certainly does not go against the regulative framework (e.g., legislation)	The principal concern of the partnership is letting involved users voice what quality they expect from the innovation	The partnership should primarily align the different goals of the involved users and the other partners	The partnership should maximally give room to the involved users to develop their own proposals for the innovation

RESULTS

After the factorial analysis, two factors remained that, explain 38 per cent of the total variance. The factors are illustrated in Table 10.2. They represent two groups of respondents who gave similar values to the statements. The distribution of the statements of each group represents a different profile on how partners think users should be involved in public-private collaborations. These values show the level of agreement with each of the roles. A positive value means that the respondents agree with the statements and a negative value means that they disagree with them. In addition, the interpretation of each statement should also be done according to the profile as a whole, not only its individual value and the pre-designated role of the statement. In the following, the main observations from the Q-methodological analyses for each of the empirical profiles are presented.

Output-oriented User Involvement

Table 10.2 shows that the statements in Profile 1 are especially aligned with the customer and partner roles as they gave a positive value, or at least not negative, to most of the statements. According to the respondents within this profile, users should be particularly involved because they can introduce alternative views and check how user-oriented the innovation is. In this viewpoint, the respondents expect the partnership to consult the users about their preferences, and to align the different perspectives of the users with each other. The

Table 10.2 Empirical profiles

	Dimension	Statement	Profile 1	Profile 2
Legitimator	Motives	Users should be involved primarily to create support for the innovation	0	0*
		Users are especially involved to check whether the rights of those they represent are guaranteed	−1	−1
	Activities	The majority of users is there predominantly to listen to what the partners have to say	−3*	−3*
		Users best leave development of innovations to others	−2*	-2*
	Role of service provider	The users should be well informed by the partnership because the innovation can then be easily accepted	1	1
		The partnership actors are there to make sure that the input of the users and other actors certainly does not go against the regulative framework (e.g., legislation)	0*	−2*
Customer	Motives	Users want to be involved primarily to indicate what they perceive as an exquisite end product	1*	0*
		Involved users should above all check how user-oriented the innovation is	2*	−1*
	Activities	Involved users have to advise the partnership about how to increase user satisfaction	1*	1*
		Just like a company asking its customers about its products, the partnership needs to consult the users about their preferences	2	2
	Role of service provider	The partnership should enable the involved users to see how the innovation works in reality	1	1
		The principal concern of the partnership is letting involved users voice what quality they expect from the innovation	0*	−1*

	Dimension	Statement	Profile 1	Profile 2
Partners	Motives	Involved users especially want to be recognized as partners	0	−1
		Users should be involved because they can have alternative views, useful for the other partners	3*	1*
	Activities	Users and the other partners should jointly define the problem and the solution	0*	3*
		Equal contributions of users and other partners (co-creation) is the only way to create relevant innovations	−1*	2*
	Role of service provider	A crucial task of the partnership is to ensure joint decision making between the involved users and the other partners	0	1
		The partnership should primarily align the different goals of the involved users and the other partners	2*	2*
Self-organizer	Motives	Users should tackle user issues themselves instead of waiting for others to do it	−2*	0*
		Users know best how to develop and organize service delivery	−2*	−1*
	Activities	Users can best define problems and solutions	−1*	−1*
		Users should set and guard the direction for the innovation process	−1*	−1*
	Role of service provider	The main role of the partnership is to provide the resources to develop proposals of the users	−1	−2
		The partnership should maximally give room to the involved users to develop their own proposals for the innovation	1*	0*

Note: * Distinguishing statements (i.e., statements that are significantly differently ranked in one factor as opposed to the other factors, with $p < 0.05$).

involved users should also advise the partnership to increase user satisfaction, while the partnerships should enable the users to see how the innovation works in practice. Users are therefore also primarily involved to indicate what they perceive as an exquisite end product. However, the partnership should also give maximum room to the involved users to develop their own ideas and should inform the users, so the innovation is more easily accepted.

On the other end of the spectrum, we find statements with which the respondents identify themselves the least. The respondents with Profile 1 disagree with most of the statements of legitimator and self-organizer roles and only gave a positive value to one statement of each of these roles. For instance, the respondents do not think that users are involved to check whether the rights of those they represent are being guaranteed. Also, they do not think that the main purpose of the partnership is to provide resources with which the users can develop their own ideas, or that co-creation is the only way to create relevant innovations. Users should also not set and guard the direction of innovation processes, and they are not best at defining problems and solutions, according to these respondents. Furthermore, the respondents indicate that users should not tackle user issues themselves, and they should also not leave the development of innovations to others. These respondents also do not think that the users know best how to develop and organize service delivery, and that the users are predominantly involved in these collaborations to listen to what the other respondents have to say.

In line with the observations indicated above, the highly ranked statements that match the 'partner' role might also have been perceived by the respondents from the perspective of output-oriented user involvement. Indeed, the respondents might view statements that point to a more active role of the users as more effective towards achieving the proper end product (e.g., 'users should be involved because they can have alternative views, useful for the other partners' and 'the partnership should primarily align the different goals of the involved users and the other partners'). Similarly, statements that were part of the legitimator role and the self-organizer role, but were highly ranked in this profile, might be necessary from the perspective of the respondents in order to obtain the required information from the users to develop the end product (e.g., 'the users should be well informed by the partnership because the innovation can then be easily accepted' and 'the partnership should maximally give room to the involved users to develop their own proposals for the innovation'). Taken together, if the users are not sufficiently informed by the partners, they will most likely also not contribute much to the end product. Similarly, if users are not enabled to develop their own proposals, it might be difficult for the partners to obtain new perspectives and create an innovative end product.

In summary, the viewpoints of the respondents in the first empirical profile are directed towards user involvement as a means to enhance the end product. Users are beneficial for the partners because they can introduce new knowledge, experiences, and perspectives which are useful for service innovation. User involvement has a functional purpose in the perspectives of these respondents, and is, hence, viewed as a means to an end.

Process-oriented User Involvement

Table 10.2 indicates that the statements in Profile 2 are particularly oriented towards the partner role, as the respondents gave a high value to most of the statements of these roles. Some statements of the customer role related to the activities of the users are also highly ranked in this empirical profile but most of the statements related to motivations and the role of the service provided have a lower value than in Profile 1. If we synthesize the statements that are highly ranked in this profile, it seems that the respondents view user involvement as a way to influence the innovation process, rather than its outcome. Indeed, the positively ranked statements refer to the way in which users can contribute during the innovation process (e.g., by contributing alternative views, advising the partnership, jointly defining the problem and solution, co-creating, joint decision making, and goal alignment). Even the statements that are aligned with the customer role might have been perceived by the respondents as a way to influence the innovation process. For instance, two-way communication activities such as advising/being consulted by the partnership is a way to actively engage in the innovation process and become part of that process. Similarly, a positive ranking of the statement 'The partnership should enable the involved users to see how the innovation works in reality', might mean that the respondents in this profile prefer to be engaged in the co-design process of the innovation.

The negatively ranked statements confirm that the respondents in Profile 2 prefer an active process role of the users in the partnership. Indeed, all the negatively ranked statements correspond to a passive role of the users, with limited responsibility. The respondents in this profile disagree that (1) the users are primarily involved in the partnership to check whether their rights are protected, (2) the users only need to listen to the partners, and leave the development of innovations to others, and (3) the partners are present in the partnership to make sure that the proposals of the users are not going against the regulative framework. These statements confirm that the respondents in this profile are not interested in a passive role for the users and that they expect a contribution from the users in the innovation process. However, they do not envision a very active user role with great responsibility in the innovation processes. Indeed, the statements that reflect the self-organizer role are negatively ranked. According to the respondents in this profile, involving users in the innovation process means that they are working together with the other partners instead of leading and directing the innovation process themselves. Hence, they value the input of the users during the innovation process but do not believe that the users should be organizing the innovation process themselves.

One statement that is negatively ranked in Profile 2 necessitates some extra explanation. The statement 'Involved users especially want to be recognized as partners' is the only statement related to the partner role that is negatively ranked. This is rather remarkable as the statement fits with the other, positively ranked statements of the partner role. However, the respondents in this profile might have interpreted the statement as a wish of the users to receive recognition or perhaps even status with their participation in the innovation process. It is clear from the negative ranking of this statement that the respondents in this profile do not believe that users are participating in the innovation process in order to be 'appointed' as real partners. The activities of the users are, according to these respondents, more important to guide the users' motives than the particular way in which they are recognized.

In summary, the respondents who adhere to the second empirical profile believe that users should be actively involved in and contribute to the innovation process. User involvement should, according to these respondents, be oriented towards the interaction and collaboration dynamics in the innovation process, rather than the expected or desired output of that process. This view on user involvement is also confirmed by the distinguishing statements, which are indicated in Table 10.2 with an asterisk.

Distribution of the Profiles

Table 10.3 illustrates the distributions of the Q-sorts over the types of actors (i.e., public/private partners and the specific types of involved organizations) and the countries of origin. The table shows that public and private partners are roughly equally distributed over the two empirical profiles. Something similar is the case for the distribution of the consultant/firms and governments. However, we can also observe that almost three times as many respondents from healthcare organizations (e.g., hospitals) adhere to Profile 1 as opposed to Profile 2. We see the opposite for the healthcare providers, of which three times as many adhere to Profile 2 in comparison to Profile 1. With regard to the countries, we observe a prominent difference between Belgium, on the one hand, in which the respondents adhere more to Profile 1, and Denmark and Estonia, on the other hand, in which the respondents adhere more to Profile 2. The respondents in the Netherlands and Spain are more evenly distributed over the two profiles than those in the other countries.

DISCUSSION AND CONCLUSION

This research shows co-existing views on the roles of user involvement in collaborative eHealth innovation projects. We obtain two profiles of collaboration partners. Respondents of both profiles agree on the active participation of users in the projects but reject the self-organizer role of users, that is, the idea of a self-governed structure with the users as initiators and where the partnership has a secondary role, particularly established to support the users in developing their ideas. The partners' views on user involvement show that they are somewhere in between viewing the users as 'customers' from a *New Public Management* paradigm and partners from the network collaboration and co-production perspective related to the *New Public Governance*.

Table 10.3 Descriptive statistics

Type of partner	**Profile 1**	**Profile 2**	**Rest**	**N**
Public partner	43.33%	33.33%	23.33%	30
Private partner	45.00%	40.00%	15.00%	20
Type of organization	**Profile 1**	**Profile 2**	**Rest**	**N**
Consultant/firm	37.50%	43.75%	18.75%	16
Government	37.50%	31.25%	31.25%	16
Healthcare organization	69.23%	23.08%	7.69%	13
Healthcare provider	20.00%	60.00%	20.00%	5
Country	**Profile 1**	**Profile 2**	**Rest**	**N**
Belgium	78.57%	7.14%	14.29%	14
Denmark	20.00%	60.00%	20.00%	5
Estonia	0.00%	60.00%	40.00%	5
Netherlands	41.67%	25.00%	33.33%	12
Spain	35.71%	57.14%	7.14%	14

Some respondents are in favour of the involvement of users, because of the potential of the users to improve the quality of the output. By ensuring that the insights of the users are included, the respondents believe that the innovation will be adopted more quickly by the users. These respondents follow the logic of New Public Management (Bryson et al., 2014; Denhardt and Denhardt, 2007), in which the consumers are motivated to participate by their individual interests as consumers, and the partnership must align the goals of the partners with those of the users to ensure that the resulting service complies with them. These respondents believe that consumer tasks should aim to provide information on consumer preferences and explain how best to meet their needs. In fact, this knowledge of consumers can improve innovation and its acceptance (Mahr et al., 2014). From an *output-oriented* perspective on user involvement, these collaboration partners view users as important contributors to the final end product of the innovation process. This role of users does not make them participants in the internal deliberation process of the partnership, so the possibilities of conflicts with the partners are lower. In this case, the effort of the leaders should focus on the inclusion of input from the consumers as the partnership may fail on this task for various reasons (i.e., lack of resources, expertise, and time or poor organizational capacity) (Olson and Bakke, 2001).

However, other respondents believe that users need to be involved because they can enhance the collaboration and innovation process itself. These respondents think that co-creation is necessary to obtain relevant innovation, which is in line with the *New Public Governance* paradigm (Brandsen and Honingh, 2016; Osborne and Strokosch, 2013; Osborne et al., 2016; Pestoff et al., 2012). These respondents also agree that users should participate by providing information about their preferences but increase the tasks on which they should be involved (i.e., in the definition of the problem and the solution). Such a *process-oriented* perspective on user involvement means that collaboration partners view users as important facilitators of the innovation process itself. This role of users means that they will be more intensively involved throughout the process, which also opens the possibility to more conflicts in the deliberation process of the project because of the discrepancies between users and partners. The involved users might have a weaker position in the partnership than the partners, and the leaders or coordinators of the partnership should empower and represent them, so their opinions are considered (Ansell and Gash, 2008).

In addition to customer-related tasks, both profiles agree on the importance of communication with the users through partnerships and showing them how the innovation works in practice. These issues may be related to the necessity of information to accept the innovation and see their individual goals achieve through their collaboration in the network (Silvia, 2011).

The profiles are not mutually exclusive as users with different characteristics (e.g., background or time availability) could have different roles in the project. Moreover, *output-oriented* user involvement can be easily instrumentalized to enable the participation of a large number of users (e.g., concept tests or satisfaction surveys) while *process-oriented* user involvement does not, as it could make the consensus-building process in the partnership challenging. Indeed, these profiles could complement each other as users involved according to *process-oriented* profiles could use the knowledge provided by the rest of the users involved to support their opinion in the decision-making process and ensure that this knowledge is correctly integrated.

The collaboration partners who support each profile show different characteristics to some extent. Healthcare organizations are more in favour of *output-oriented* user involvement while healthcare providers align more with *process-oriented.* Healthcare organizations may have a more comprehensive view of all healthcare processes and doubt the ability of users to consider all factors related to innovation (e.g., financial sustainability). In contrast, healthcare professionals often have no management responsibilities and, as such, are not responsible for overarching management decisions that affect the whole organization, so they may be more inclined to give users a larger role. Additionally, these different partners' viewpoints might be explained by differences in culture, institutions, or policy between countries. Respondents from Belgium, which has a Napoleonic administrative tradition (Pollitt and Bouckaert, 2017), are less inclined to citizen participation and are more aligned with *output-oriented* user involvement. Indeed, Belgian public organiuzations collaborate in innovation processes far more frequently with other governments or private sector organizations than with citizens and users (Verhoest et al., 2020). On the other hand, most respondents from Denmark, which has a stronger tradition of engaging citizens (Pollitt and Bouckaert, 2017), support *process-oriented* user involvement. The case of the respondents from Estonia could be explained by the projects' and partners' specific characteristics more than by country-level characteristics.

The differences between partners' viewpoints on user involvement could lead to role-related conflicts (Mele, 2011). In such cases, consensus building might be more challenging and will need the assistance of strong leaders in the project (Ansell and Gash, 2008). Hence, besides identifying and incorporating the right members for the partnership, the project coordinator should consider the different users needed depending on the vision of the partners, to avoid any misalignment between the viewpoints on user involvement of the partners.

Conflicts could also emerge between the users involved and the partners. This research only considers partners' viewpoints and further research could compare users' and partners' viewpoints on user involvement, in order to reveal possible conflicts regarding the role expected by users and what the partners are willing to accept. This research shows the co-existence of two public administration styles regarding user involvement (New Public Management and New Public Governance) but discards other considerations of user involvement. The 'self' in the concept of self-governance concept includes a wide variety of public and private actors (Sørensen and Triantafillou, 2009). However, the rejection of partners to consider users as initiators or leaders of the innovation process shows a limit to user involvement in self-governed structures. This restriction might be because of the advanced knowledge requirements in information and communication technology and health, necessary for the development of eHealth innovations and the complexity of the healthcare sector. Further research could extend this research to other fields to test whether partners' views are common or depend on the field of the innovation project.

NOTES

1. The TROPICO project received funding from the European Union's Horizon 2020 research and innovation programme under grant agreement No. 726840. For more information: https://cordis.europa.eu/project/id/726840
2. Although Q-methodology requires a proportional number of respondents for a given number of statements (most often a 1:1 ratio) (Watts and Stenner, 2012), we chose to include more respondents because of the inherent variance in our research design (multiple countries, multiple types of actors in the partnerships, multiple types of eHealth services). A small P-set might have been insufficient to capture the large variety of the projects. Furthermore, studies conducted in

multiple countries generally consider larger P-sets, and manage to obtain valid results (e.g., ratio of 5:1 in Warsen et al., 2019).

3. The Q-sort data was analysed using the Q-methodology package KenQ, which calculates the correlation matrix, the eigenvalues, and the factor loadings from the data. Three cumulative measures are important to decide which factors are strong enough to be retained. First, we used the Kaiser-Gutmann criterion, which proposes that factors should have an eigenvalue equal or greater than 1 before they are retained (Watts and Stenner, 2012). Second, we only retained the factors which possessed at least two factor loadings which were statistically significant (calculated by $1.96 \times \frac{1}{\sqrt{Number\ of\ items}}$, $p < 0.05$; Watts and Stenner, 2012). Third, a factor was only retained when the explained variance of the factor was at least 7 per cent and the cumulative variance of the selected factors was larger than 30 per cent (Molenveld et al., 2019). Additionally, varimax factor rotation was used to interpret the factors correctly (Watts and Stenner, 2012).

REFERENCES

Andreassen, H.K., Kjekshus, L.E., and Tjora, A. (2015). Survival of the project: A case study of ICT innovation in health care. *Social Science and Medicine, 132*, 62–9.

Ansell, C., and Gash, A. (2008). Collaborative governance in theory and practice. *Journal of Public Administration Research and Theory, 18*(4), 543–71.

Brandsen, T., and Honingh, M. (2016). Distinguishing different types of coproduction: A conceptual analysis based on the classical definitions. *Public Administration Review, 76*(3), 427–35.

Bryson, J.M., Crosby, B.C., and Bloomberg, L. (2014). Public value governance: Moving beyond traditional public administration and the new public management. Public Administration Review, 74(4), 445–56.

Burt, C., and Stephenson, W. (1939). Alternative views on correlations between persons. *Psychometrika, 4*(4), 269–81.

Callens, C., Verhoest, K., Klijn, E.H., Nõmmik, S., Pina, V., and Brogaard, L. (2023). How service users envision their engagement in processes of collaborative innovation: A Q-methodological study on user involvement in eHealth collaborations. *Public Policy and Administration.*

Daya, I., Hamilton, B., and Roper, C. (2019). Authentic engagement: A conceptual model for welcoming diverse and challenging consumer and survivor views in mental health research, policy, and practice. *International Journal of Mental Health Nursing, 29*(2), 299–311.

Denhardt, J.V., and Denhardt, R.B. (2007). *The New Public Service: Serving, Not Steering*, New York: M.E. Sharpe.

Dryzek, J.S., and Berejikian, J. (1993). Reconstructive democratic theory. *The American Political Science Review, 87*(1), 48–60.

European Commission (2018). *Communication from the Commission to the European Parliament, the Council, the European Economic and Social Committee and the Committee of the Regions on enabling the digital transformation of health and care in the Digital Single Market; empowering citizens and building a healthier society (25/4/2018).* https://eur-lex.europa.eu/legal-content/EN/TXT/?uri=COM%3A2018%3A233%3AFIN

Gremyr, I., Elg, M., Smith, F., and Gustavsson, S. (2018). Exploring the phase for highest impact on radicality: A cross-sectional study of patient involvement in quality improvement in Swedish healthcare. *BMJ Open,* 8, e021958.

Jang-Jaccard, J., Nepal, S., Alem, L., and Li, J. (2014), Barriers for delivering telehealth in rural Australia: A review based on Australian trials and studies. *Telemedicine and E-Health, 20*(5), 496–504.

Klijn, E.H., and Koppenjan, J. (2012). Governance network theory: Past, present and future. *Policy & Politics, 40*(4), 587–606.

Koppenjan, J., and Koliba, C. (2013). Transformations towards new public governance: Can the new paradigm handle complexity? *International Review of Public Administration, 18*(2), 1–8.

Mahr, D., Lievens, A., and Blazevic, V. (2014). The value of customer co-created knowledge during the innovation process. *Journal of Product Innovation Management, 31*(3), 599–615.

McGuire, M. (2006). Collaborative public management: Assessing what we know and how we know it. *Public Administration Review, 66*(s1), 33–43.

Mele, C. (2011). Conflicts and value co-creation in project networks. *Industrial Marketing Management, 40*(8), 1377–85.

Molenveld, A., Verhoest, K., Voets, J. and Steen, T. 2019. Images of coordination: How implementing organisations perceive coordination arrangements. *Public Administration Review, 80*(1), 9–22.

Nederhand, J., Bekkers, V., and Voorberg, W. (2016). Self-organization and the role of government: How and why does self-organization evolve in the shadow of hierarchy? *Public Management Review, 18*(7), 1063–84.

Nederhand, J., Klijn, E.H. van der Steen, M., and van Twist, M. (2019). The governance of self-organization: Which governance strategy do policy officials and citizens prefer? *Policy Sciences, 52*(2), 233–53.

Olson, E.L., and Bakke, G. (2001). Implementing the lead user method in a high technology firm: A longitudinal study of intentions versus actions. *Journal of Product Innovation Management, 18*(6), 388–95.

Osborne, S. (2010). *The New Public Governance? Emerging Perspectives on the Theory and Practice of Public Governance*. London: Routledge.

Osborne, S.P., and Strokosch, K. (2013). It takes two to Tango? Understanding the co-production of public services by integrating the services management and public administration perspectives. *British Journal of Management, 24*(s1), S31–S47.

Osborne, S.P., Radnor, Z., and Strokosch, K. (2016). Co-production and the co-creation of value in public services: A suitable case for treatment? *Public Management Review, 18*(5), 639–53.

Pestoff, V., Brandsen, T., and Verschuere, B. (2012). *New Public Governance, the Third Sector and Co-production*. London: Routledge.

Pierre, B.G., and Peters, J. (2020). *Governance, Politics and the State*. New York: Bloomsbury Publishing.

Pollitt, C., and Bouckaert, G. (2017). *Public Management Reform: A Comparative Analysis – into the Age of Austerity*. Oxford: Oxford University Press.

Sangill, C., Buus, N., Hybholt, L., and Berring, L.L. (2019). Service user's actual involvement in mental health research practices – a scoping review. *International Journal of Mental Health Nursing, 28*(4), 798–815.

Silvia, C. (2011). Collaborative governance concepts for successful network leadership. *State and Local Government Review, 43*(1), 66–71.

Sørensen, E., and Torfing, J. (2009). Making governance networks effective and democratic through metagovernance. *Public Administration, 87*(2), 234–58.
Sørensen, E., and Torfing, J. (2011). Enhancing collaborative innovation in the public sector. *Administration & Society, 43*(8), 842–68.
Sørensen, E., and Triantafillou, P. (2009). *The Politics of Self-governance*. London: Routledge.
Torfing, J. (2019). Collaborative innovation in the public sector: The argument, *Public Management Review, 21*(1), 1–11.
Torfing, J., and Triantafillou, P. (2013). What's in a name? Grasping new public governance as a political-administrative system. *International Review of Public Administration, 18*(2), 9–25.
Torfing, J., Andersen, L.B., Greve, C., and Klausen, K.K. (2020). *Public Governance Paradigms: Competing and Cno-existing*. Cheltenham, UK and Northampton, MA, USA: Edward Elgar Publishing.
Urueña, A., Hidalgo, A., and Arenas, Á.E. (2016). Identifying capabilities in innovation projects: Evidences from eHealth. *Journal of Business Research, 69*(11), 4843–8.
Van Eijk, C.J., and Steen, T.P. (2014). Why people co-produce: Analysing citizens' perceptions on co-planning engagement in health care services. *Public Management Review, 16*(3), 358–82.
Van Eijk, C., Steen, T., and Verschuere, B. (2017). Co-producing safety in the local community: A Q-methodology study on the incentives of Belgian and Dutch members of neighbourhood watch schemes. *Local Government Studies, 43*(3), 323–43.
Van Exel, J., and De Graaf, G. (2005). *Q Methodology: A Sneak Preview*. https://qmethod.org/
Vangen, S., and Huxham, C. (2003). Nurturing collaborative relations: Building trust in interorganizational collaboration. *The Journal of Applied Behavioral Science, 39*(1), 5–31.
Verhoest, K., Steen, T., Langbroek, T., Van Dijck, C., Dockx, E., Riche, C., and Lingier, P. (2020). *Gap Analysis on Capacities for Collaborative Innovation in the Belgian Federal Administration: Survey Outline, Descriptive and Explanatory Analyses (WP6). Report as part of the BRAIN-funded 'Public Sector Innovation through Collaboration' project (PSICO)*. Antwerpen: Universiteit Antwerpen.
Von Hippel, E. (1986). Lead users: A source of novel product concepts. *Management Science, 32*(7), 791–805.
Von Hippel, E.A., Ogawa, S., and De Jong, P.J. (2011). The age of the consumer-innovator. *MIT Sloan Management Review, 53*(1), 27–35.
Voorberg, W., Bekkers, V., and Tummers, L. (2014). *Co-creation in Social Innovation: A Comparative Case-Study on the Influential Factors and Outcomes of Co-creation*. IRSPM, Ottawa, Canada.
Watts, S., and Stenner, P. (2012). *Doing Q Methodological Research Theory, Method & Interpretation*. London: Sage.
Warsen, R., Klijn, E.H., and Koppenjan, J. (2019). Mix and match: How contractual and relational conditions are combined in successful public–private partnerships. *Journal of Public Administration Research and Theory, 29*(3), 375–93.

11. It is all about interaction: network structure, actor importance, and the relation to innovative outcomes

Tom Langbroek and Koen Verhoest

INTRODUCTION

In recent years, many governments have been establishing collaborative arrangements to develop public sector innovations to cope with today's societal problems (Hartley et al., 2013; Lopes & Farias, 2020; Torfing, 2019). Public sector innovation entails the development and implementation of new public policies, services, technologies, and administrative processes that represent a qualitative change from how things were done before (De Vries et al., 2016; Gieske et al., 2019). Although working in inter-organisational collaborative arrangements is not a new phenomenon, public sector innovations have been increasingly developed in collaborative arrangements in recent years. New opportunities to interact have arisen in the past decades, such as digital tools that allow interaction with actors that would otherwise be left out of the innovation process (Castells, 2000; Geuijen et al., 2017). But also, the increasing fragmentation of society and the subsequent interdependencies between actors have led to a need for inter-organisational collaboration (Agger & Lund, 2017; Bommert, 2010).

A concept closely related to interdependencies in inter-organisational collaboration is that of actor importance (Koppenjan & Klijn, 2004:178; Meijer, 2014). Stevens (2018) is one of the few scholars who examined actor importance in collaborative arrangements aimed at innovation in relation to network structure. He found that individual actors are in some cases more likely to interact with actors they find 'very necessary' to tackle the policy problem.

Recently, considerable research has examined the relationship between network structure and network effectiveness (e.g., Cepiku et al., 2020; Raab et al., 2015; Stevens, 2018). Research suggests that network structure can positively influence the outcomes of the collaboration. For example, a high degree of network-level connectedness allows information to flow efficiently through the network and is associated with the development of social capital and trust (Bodin et al., 2017; Hu et al., 2022; Yi, 2018).

Clique overlap is named the most effective way type of network integration by milestone studies such as those by Provan and Milward (1995) and Provan and Sebastian (1998). A clique is a group of at least three actors who are directly connected with each other. Cliques overlap when an actor is a member of multiple cliques and thus connects multiple cliques with each other. When clique overlap occurs, actors in the network are more closely connected to each other, while superfluous ties between actors are reduced. Remarkably, little attention has been paid to how innovative outcomes can be explained by the integration through clique overlap of the most important actors. While the inclusion of important actors who are necessary for the realisation of the innovation is at the basis of collaborative innovation (Torfing, 2019), little is known about the extent to which the integration of important actors in collaborative arrangements results in innovative outcomes.

Therefore, this study answers the following research question: *How can innovative outcomes of collaborative public sector innovation projects be explained by the network integration of its most important actors?*

The question is answered by examining three cases in which a collaborative arrangement was established in order to create better digital information exchange in the public sector. For each of these cases, the degree of clique formation and clique overlap, and the resulting network integration of the most important actors are examined in two interaction networks: interaction outside meetings and interaction during meetings.

THEORETICAL FRAMEWORK

Collaborative Innovation and the Importance of Actors

Governments increasingly turn to the development of public sector innovations as a way to deal with wicked problems that arise from a complex, fragmented society, unpredictable events, and increasing citizens' demands for public services (Ansell et al., 2021; Lopes & Farias, 2020; Torfing, 2019; Wegrich, 2019). Although research on innovation has gained increasing attention, no clear consensus on the definition of the concept is present. Following the literature review by De Vries and colleagues (2016) definitions of public sector innovation commonly emphasise innovations as being something new within a given context. This can be a new or changed service, but also a new policy, technology, process, etc. This novelty might exist somewhere else but is new in its context and represents a change and discontinuity with how things were done before (Damanpour et al., 2009; Gieske et al., 2019; Osborne & Brown, 2011). Innovation is, therefore, something different from optimisation in the sense that innovation represents a break from the past and concerns the implementation of something new in the context. Optimisation is, on the other hand, an improvement of existing routines in line with the past. In recent studies, one additional characteristic of public innovation is commonly acknowledged. Unlike private sector innovation, which is driven by competitive advantage over others, public sector innovation is specifically aimed at the creation of public value instead. The innovation aims to solve a societal problem and adds value to society in that way (Chen et al., 2020; Crosby et al., 2017).

As all necessary resources for the development of innovations are usually not available within one single government organisation, governments increasingly develop public sector innovations in collaborative arrangements (Bommert, 2010; Torfing, 2019). In these collaborative arrangements resources such as knowledge, financial means, and production resources are combined to develop an innovation to cope with the problem at hand. Still, merely bringing actors together does not result in innovations. Several factors are closely linked to the process between collaboration, on the one hand, and the creation of innovation, on the other hand. Among them are:

(1) The inclusion of the necessary actors in the collaborative arrangement and their importance (Ansell & Torfing, 2014:11; Godenhjelm & Johanson, 2018; Siddiki et al., 2017). For example, actors with knowledge, financial means, decision-making power, etc.
(2) The interactions among these actors (Agger & Sørensen, 2018; Lewis et al., 2018; Lopes & Farias, 2020). For example, interaction among people with different insights or knowledge allows actors to learn and spurs the generating of new ideas (Koebele, 2019; Voorberg et al., 2017).

Actor Inclusion, importance, and Interaction

Concerning the inclusion of actors, actors are included for different reasons in the innovation process (Godenhjelm & Johanson, 2018). These can be the resources these actors can bring to the process, such as different insights or financial resources (see, e.g., resource dependency theory; Hillman et al., 2009; Pfeffer & Salancik, 1978; Scharpf, 1978), the coordinating role the actor fulfils, their decision-making power, or simply because they are interested in the problem at hand and want to think along (Koppenjan & Klijn, 2004:178). Actors become more important as the substitutability of, for example, their knowledge or decision-making power is low. Therefore, a low substitutability of necessary resources is a basis of power in collaborative arrangements aimed at innovation and thus can make certain actors more important than others.

Next, the premise of collaborative innovation is to interact with each other in order to combine different resources and perspectives, learn from each other, and subsequently implement the innovation (Ansell & Torfing, 2014:11). Information sharing is crucial for the development of innovations (Koliba et al., 2017). During meetings, actors need to build upon each other's ideas to deepen discussions, come to a synergetic process, and learn from each other. Outside meetings, information sharing is necessary to elaborate on the things discussed during official meetings, work out details etc.

The structure of the interaction in these collaborative arrangements (also referred to as networks) can take many forms and shapes. Network characteristics such as the density, centrality of individual actors, and structural holes reflect the shape of the network, and thus the interaction patterns within the collaborative arrangements (Lusher et al., 2012:7). Individuals gain access to information, social support, and other resources through the ties with other actors (Agneessens & Wittek, 2012; Bodin et al., 2017; Hu et al., 2022; Yi, 2018). Moreover, research suggests that central actors are more likely to access useful knowledge from others and therefore become more important (Tsai, 2001; Zhao, 2022). They are the 'spiders in the web'. Hence, the causal relation between importance and network integration is somewhat unclear as importance might not only be attributed to actor-specific characteristics such as the possession of resources, but also to an actor's central network position.

Milestone studies in the relation between network structure and network outcomes are the studies of Provan and Milward (1995) and Provan and Sebastian (1998), who provided a framework for the determinants of effective network outcomes; one of them being the integration of actors in the network. Three types of integration are commonly distinguished in the inter-organisational network literature. First, density-based integration, which is the type of integration based on the total level of ties among the actors in the network (Scott, 2000:69). In this type of integration the observed number of ties between the actors is compared to the maximum number of ties. A higher density resembles a higher degree of network integration. A second type of integration is centralised integration, which is the extent to which the network ties are organised around particular focal actors (Borgatti et al., 2013:149). The third type of commonly distinguished type of integration is clique overlap (Borgatti et al., 2013:184; Raab et al., 2015). Cliques are a minimum of three different actors who are directly connected to each other in a network. When actors are a member of multiple cliques it results in clique overlap. In the case of clique overlap, an actor connects multiple cliques with each other and thus indirectly connects actors who are not in the same clique.

The discussion has focused on what kind of integration is best for effective network outcomes and is commonly focused on the balance between density-based integration and centralised integration (Provan & Sebastian,

1998; Saz-Carranza & Ospina, 2011; Ngamassi et al., 2014; Turrini et al., 2010). Several studies found that centralised integration organised around a central coordinator is positively associated with network outcomes (Cristofoli & Markovic, 2016; Provan & Milward, 1995; Raab et al., 2015). Free-wheeling behaviour in the network can be prevented as the central coordinator is able to control and monitor the behaviour of all the other actors, especially in sparsely connected networks (Cristofoli et al., 2021).

However, in this study, we examine integration through clique overlap which is found to combine the advantages of strong density-based integration and strong centrality-based integration and enhances the overall network outcomes but is very overlooked in current literature (Provan & Sebastian, 1998). The membership of a clique in the network has been linked to several benefits, such as a higher pace of information sharing through the network and facilitation of learning (Provan & Sebastian, 1998). It was found that network outcomes are evaluated more positively when cliques overlap each other, but less positively when too many actors in the network are connected to each other. Having clique overlap is effective because members who are in several cliques are brokers and thus connect different cliques to each other. In this structure, not everyone in the network needs to be directly connected to each other because the information is transferred through the actors that are members of multiple cliques. Moreover, while a certain amount of dispersion of important actors throughout the network seems needed, (complete) separation of certain actors is found to be detrimental to successful outcomes (Yi, 2018). Stevens (2018) examined how actor importance determines interaction in collaborative innovation networks that work towards a joint outcome and found that actors are in some cases more likely to interact with actors they find 'very necessary' to tackle the policy problem. This suggests that actors who find each other important tend to stick to each other. A core group of important actors that mainly directly interact with each other without interacting with others in the network would be an obstacle to the development of innovations (Provan & Kenis, 2008).

From these previous findings we expect that cases with successful innovative outcomes are characterised by a high degree of clique overlap in terms of the most important actors being included in multiple cliques, both during and outside meetings.

RESEARCH DESIGN AND MEASUREMENT

Cases and Data Collection

Three Belgian cases were examined in which a collaborative arrangement was created with the goal to implement a public sector innovation. These innovations aimed to create a better digital information exchange between the key players involved in the policy field at hand. They all focus on a more effective way to cope with information needs by transforming information systems in order to work more effectively. They are clear examples of the line of reasoning that effective information-sharing systems are a lynchpin in critical public policy areas to be effective and that government must embrace the digital era to optimise inter-organisational information integration (Meijer and Bekkers, 2015; Pardo and Tayi, 2007).

All actors in the network were asked to participate in a survey that asked about their interactions with, and their perception of the importance of the other actors.[1] The cases were selected based on the following criteria: (1) The cases entail arrangements involving public actors and to the extent possible also private actors and citizens; (2) In order to avoid the pro-innovation bias we included also cases which did not materialise in innovations, or in which innovation processes were particularly difficult in their progress; (3) Comparability in terms of network size, goal of the innovation (creating new digital procedures to solve difficulties in governmental processes), and accessibility were important criteria as well.

CareLab was a project initiated by the Belgian federal government focusing on the simplification of rules and bureaucracy for parents with a disabled child. A core group of 18 actors, such as health professionals, civil servants, and parents could be identified. The project ended with the idea selection of four innovations, including a digital government tool to reduce the administrative burden for parents with a disabled child, and a first step towards implementation. After that, core actors left the process and sustainable implementation of these solutions did not take place.

Invasive Species was a project to generate a more comprehensive and effective policy on invasive species by creating a new institutional arrangement that organises and formalises digital information exchange between institutions dealing with invasive species policies across Belgian regions and communities. A core group of 11 actors could be identified. These included federal and regional policy officers, scientists, and legal experts.

Radicalisation was a process innovation with the goal to change the digital information exchange procedure concerning signs of radicalisation within the group of asylum seekers or refugees to ensure that the transfer of information on radicalism is effective, both horizontally and vertically. A new (digital) notification procedure to detect radicalisation and the new way of information exchange was implemented. A core group of ten actors, such as representatives of the Federal Agency for the Reception of Asylum Seekers and the General Intelligence and Security Service could be identified.

Interaction

Two types of interaction networks were studied in each case. To capture interaction in the collaborative arrangement outside official meetings we looked at the network of information sharing *outside official meetings*. Moreover, we also looked at the network concerning elaboration upon others' ideas *during official meetings* to determine which actors interact with each other during meetings. We will refer to these networks as *interaction outside meetings* and *interaction during meetings*.

To map the interactions outside formal meetings, the following question was asked:

- *Could you please indicate to whom you gave and from whom you received information, after and outside of formal meetings? 'Information' includes reports, statistics, advice, and remarks. This information can be both verbal and written.*

A tie between actors was only considered when it was confirmed from both sides. For example, if actor 'i' claimed that he gave information to actor 'j', it was only regarded a tie when actor 'j' indicated that he received information from actor 'i'. When this was not possible due to missing network data because an actor did not fill out the survey (CareLab: 2 actors, Invasive Species: 2 actors, Radicalisation: 1 actor), a tie was considered by confirmation from only one respondent.

To determine the interaction *during meetings*, we asked:

- *Which participants in [project name] most frequently elaborated during the meetings of [the arrangement] upon the information and ideas you shared?*

For this question, respondents could indicate the five participants that elaborated most frequently on their contributions inside the meetings. Because CareLab consisted of more actors, the respondents could name up to eight actors in this case to make the networks comparable. That way respondents could name around 50 per cent of the actors in all cases. We decided to pose this question in a way that respondents were not able to name every actor in the collaborative arrangement. As the collaborative arrangements consisted of relatively few actors, we were only interested in the actors who elaborated most on an actor's contributions. The meetings in the cases were set up in a way that all actors engaged in group discussions with each other. By limiting the number of actors, we prevented respondents from naming every actor in the collaborative arrangement.

Actor Importance

Respondents were asked to evaluate how important they perceived the other actors in order for the innovation process to succeed. They were asked to answer the question:

- *Could you please indicate for each of the participating actors whether you considered it 'very important', 'important' or 'not that important' that they were involved in the process.*

We took only the 'very important' answer category into account as respondents were not likely to indicate actors as 'being unimportant for the process'. By only including the 'very important' category, we got a clear view of the actors respondents felt were the most important, as it eliminated the rather neutral 'important' category, and, thus, giving a better representation of the most important actors.

The importance of the individual actors in the networks was determined by calculating the percentage of the times they were labelled as being 'very important' by the other actors. For example, in CareLab, the total number of times an actor was reported to be 'very important' to the process was divided by the total number of actors minus three (as actors cannot indicate themselves as being very important and the presence of two non-respondents in this case).

Network Integration

Multiple measures were used to determine the level of integration of each network. As a basic variable, the density of the networks was checked to see how the actors in the whole network are connected to each other. The density is the number of ties between actors compared to the maximum number of ties (Borgatti et al., 2013:183). Density is used as a measure of network integration, as is common in studies examining network structures (Ngamassi et al., 2014; Provan & Sebastian, 1998).

Next, we examined the different cliques and clique overlap in the networks. As mentioned, a clique is a subset of at least three actors in which every actor is adjacent to every actor in the subset, and it is impossible to add any more actors to this clique without violating this condition (Borgatti et al., 2013:183). As we are interested in the clique formation of interactions, we only looked at reciprocal ties. We excluded cliques of interaction in which, for example, only one actor gives information and the other actors in the clique 'just' receive information. That way, we only took cliques into account in which all actors actively gave information to and received information from all other actors in the clique. This principle was applied to both interaction networks.

The examination of cliques enabled us to identify central actors, and/or whether (groups of) actors are totally isolated. Because the networks are relatively small, we looked for cliques in which all actors of the clique were directly connected to each other. Therefore, methods of analysis that 'loosen' the strict definition of a clique by allowing actors to be not necessarily connected to every other actor in the clique, such as N-clans (Harary et al., 1965) or K-plexes (Seidman & Foster, 1978) were not used. The cliques were identified using the UCINET software. With this software several key measures of network integration were obtained (Kegen, 2015):

- The number of cliques.
- The clique density (the number of actors in at least one clique).
- The average size of cliques.
- The individual and average clique centrality. This is the absolute number of cliques an individual actor is a member of and, at a network level, the average amount of cliques an actor is a member of.
- The integration of clusters of cliques. The cliques were analysed using an average link hierarchical clustering procedure to see how cliques overlap with each other (Borgatti et al., 2013:96). Average linkage hierarchical clustering is a stepwise procedure for determining the clusters in the network based on the average distance from any member of one clique to any member of the other cliques. The algorithm merges the closest pairs of cliques into a cluster. Then, the clique that is closest to this new cluster is, in turn, merged with this cluster, etc. This procedure is repeated until all cliques are merged into a single cluster. As high clique overlap in the network requires less stages of this so-called clustering, the lower the level (stages) of clustering, the higher the extent of clique overlap.

Innovative Outcome Measures

The innovative outcomes of the cases were determined in two ways. First, every respondent was asked to rate the innovative outcomes of the project on a scale ranging from 0 to 10 using four items.[2] Once data were collected, the results of the three cases were pooled and a factor analysis (principal components analysis) was executed to come to one broad measure of innovative outcomes. Second, to overcome a possible bias in the respondents' answers, the phase at which the project ended was determined based on interview data and official documents. The commonly used phases of the innovation cycle – idea generation, idea selection, implementation, and dissemination – were applied to the cases (Sørensen & Torfing, 2011).

RESULTS

Concerning the innovative outcomes of the cases, the factor analysis indicated that items loaded on one factor and the scale was regarded as reliable (Cronbach's alpha = 0.733), the mean factor scores were then calculated to obtain comparable measurements per case (see also Provan & Sebastian, 1998). The results show that CareLab scores substantially lower on innovative outcomes than the other projects, with the negative factor score (−0.34) indicating that the project was ineffective in producing innovative outcomes. The process of the Invasive Species (factor score: 0.20) and Radicalisation case (factor score: 0.46) was much more successful in producing innovative outcomes.

As CareLab did not move past the idea selection phase, it is possible to say that this project was less successful than the other two cases as these reached the implementation phase.

We expect that cases with successful innovative outcomes are characterised by a high degree of clique overlap in terms of the most important actors being included in multiple cliques, both during and outside meetings. The remainder of this section presents the two types of networks separately. The networks are complementary to each other, but it makes more sense to look at each network separately as the interactions in the networks are different in nature. During meetings, actors build upon each other's ideas to deepen discussions and to come to a synergetic process. Outside meetings, actors can elaborate on the things discussed during official meetings and work out the details in smaller groups. By comparing the networks separately, it is easier to notice the differences between the cases per type of interaction network and therefore what the characteristics of a successful case are compared to a less successful case and how it is different per type of interaction network. Moreover, the networks are measured in different ways, which makes a separate presentation more suitable.

Interaction Network 1: Interaction Outside Meetings

Table 11.1 shows the integration of the 'interaction outside meetings' networks. Comparing the interaction outside meetings network across the three cases, CareLab shows the lowest density, followed by the Invasive Species case, and Radicalisation has the highest network density. This can partly be explained as the relative measure of density typically decreases when network size increases (Jansen, 2006:194). An actor is only able to have a direct tie with a limited number of other actors so when the network size increases, the relative number of linkages decreases. Larger network size is generally associated with a higher number of cliques, because there are simply more actors to connect to. This is visible in the network data as CareLab has one clique more than the successful Invasive Species and Radicalisation networks. However, the clique density of the CareLab network can be considered quite low as just a little over half of the actors (10 out 18 actors) is integrated in at least one clique.

Clique density is a measure that indicates how many actors are a member of at least one clique. The clique density is lowest in the CareLab network, and higher in the smaller Invasive Species and Radicalisation networks. This is in line with the argument that smaller networks have a higher clique density as isolated actors are spotted more easily and are consequently sooner squeezed into a clique (Kegen, 2015). Furthermore, the cliques in the CareLab network all contain the minimum number of three actors, while the other networks tend to have some cliques containing four actors, which gives these networks a higher average clique size pointing to a higher extent of integration.

Clique Overlap

The clique centralities were obtained to spot indications of clique overlap. The clique centrality indicates the number of cliques an actor is a member of. Moreover, a hierarchical clustering procedure was executed to see to what extent cliques are integrated with each other. The average clique centrality of CareLab is 1, indicating that on average every actor is a member of one clique. However, as the clique density indicates, only 55 per cent of the actors are present in at least one clique. The average clique centrality turned out to be especially high because one actor is member of all six cliques, and one is a member of four cliques. This indicates a high extent of clique overlap but given the observation that only 55 per cent of the actors is included in a clique it means that especially a core group is well connected to each other through clique overlap and the other actors are more isolated (eight of them are in no clique at all). In this network it indicates a strong centralisation towards a core group that is closely tied together, while other actors are more isolated.

Actors in the more successful Invasive Species and Radicalisation networks are less isolated. The average clique centralities indicate that actors are on average a member of 1.55 cliques (Invasive Species) and 1.60 cliques (Radicalisation). An interesting observation in the Radicalisation network is that one actor is present in all cliques in the network, indicating strong clique overlap through this central actor, just as in the CareLab network. Not surprisingly, this central actor is the coordinator.

However, although a central coordinator is present with whom actors are directly connected in the Radicalisation network, the observation that a higher percentage of actors in the network are a member of at least one clique (70 per

Table 11.1 *Network integration 'interaction outside meetings' network*

	CareLab	Invasive Species	Radicalisation
Network density	0.127	0.273	0.311
No. of cliques in network	6	5	5
Average clique size	3	3.4	3.2
Clique density	0.55	0.72	0.70
Average clique centrality	1	1.55	1.60
Individual clique centralities	0 to 6	0 to 3	0 to 5
Complete integration of cliques at level …	2	3	2
Top most important actors* (with individual clique centrality)	1. Local coordinator 1: 93% (clique centrality: 6) 2. Parent 1: 67% (0) 3. Private actor 5: 60% 4. Federal coordinator: 60% (4)	1. Public actor 1 Federal: 50% (0) 2. Public actor 2 Flemish 38% (1) 3. Public actor 2 Walloon: 25% (2) 4. Public actor 1 Brussels: 25% (0)	1. Federal actor 4 (coordinator) 78% (5) 2. Federal actor 1 67% (3) 3. Federal actor 2 67% (2)

Note: * Top three. In case of equal importance more actors are listed in the table.

cent) indicates a high degree of interconnectedness of the other actors as well. Therefore, in this network actors interact frequently with the coordinator, but unlike CareLab, they also interact with the other actors.

In the Invasive Species network, none of the actors are a member of every clique, meaning that no central actor connects all cliques with each other. In this network, 72 per cent of the actors are in either one, two, or three of the five different cliques. Not having a central actor that connects all cliques implies less clique overlap, which is also confirmed by the lowest level of clique integration following the hierarchical clustering procedure. However, as only 28 per cent of actors are in no clique at all and cliques are on average larger, we are able to say that actors are more directly connected to each other than in CareLab where interaction concentrates towards a well-connected core group.

Actor Importance

Then, concerning actor importance, especially the CareLab and Radicalisation cases have actors who are regarded to be very important to the process by the vast majority of the collaborative arrangement. The most important actor in these cases are coordinators and are a member of all cliques in the 'interaction outside meeting' network, indicating that the most important actor in the network is well integrated with the cliques in these 'interactions outside meetings' networks. However, the other most important actors in the CareLab network, not being coordinators, are less well integrated into the network. These important actors are in only one or even none of the cliques. The important actors in the Radicalisation case are in that respect better integrated into the network as they are all a member of multiple cliques. Interestingly, actors in the Invasive Species case tend to qualify other actors in the network less frequently as being very important to the process. The most important actor is only named 'very important' by half of the other actors. Also, the most important actors, in this case, are poorly integrated into cliques. Especially less important actors are well integrated into the clique structure of the network.

Following from these findings we can conclude that the CareLab network has the lowest level of integration on the whole network level with a core group of actors that interact with each other, while other actors, including some of the most important ones, are in the periphery of the network and hardly share information with each other outside meetings. The cliques overlap to a high extent, but only a few actors are present in these cliques, making the actors who are a member of a clique well connected with each other, but poorly with the rest of the (sometimes highly important regarded) actors. This points to a very important regarded 'in-group' that mainly shares information with each other, while the other less important actors hardly interact with each other.

Clique overlap is lower in the Invasive Species network, but as more actors are present in at least one clique, the integration of the whole network of Invasive Species is higher. Also, important actors are more dispersed throughout the network; none of the important actors are present in all cliques, which means that the important actors have a less prominent role in this network.

The Radicalisation network shows a high level of integration as a majority of actors are included in a clique and this network has a high level of clique overlap. Because a vast majority of actors are a member of a clique and cliques overlap to a high extent, the whole network is tightly connected. Especially the most important actors are well integrated in the network having a membership of multiple cliques, which means that important actors are at the core of the network, yet they are better connected to the others in the network unlike what we see in the CareLab network.

Interaction Network 2: Interaction During Meetings

The 'interaction during meetings' networks shows more extreme results between the different cases. As presented in Table 11.2, the unsuccessful

CareLab case has a very low overall density, meaning that actors are to a low extent connected to each other. Especially actors in the Invasive Species case tend to elaborate on each other's ideas as almost half of the actors are directly connected to each other.

Clique Overlap

Especially the actors in the Invasive Species network are highly integrated as everyone is part of at least one clique in the network, meaning that every respondent is part of a group of a minimum of three actors who frequently elaborate upon each other's ideas. We must stress, however, that the two non-respondents, in this case, were not taken into account.[3]

Whereas in the Invasive Species network, every actor is a member of at least one of the five different cliques making this a well-integrated network, the other networks contain fewer cliques and only a small percentage of the actors are members of at least one clique. In the Radicalisation network, only three actors form a clique together, and in the CareLab network just four actors.

This means that in both a successful and an unsuccessful case only a small core group of actors elaborated upon each other's ideas. The Invasive Species network consists of substantially larger cliques than the other networks with an average size of 4.6 compared to the minimum amount of three actors in the other cases, indicating strong integration of actors in this network. The high average clique centrality indicates that actors are on average a member of 2.54 cliques, which is considerably higher than in the other cases.

The individual clique centrality shows that, again, no actor in the Invasive Species case is a member of all five cliques. However, multiple actors have an individual clique centrality of four, indicating membership of four cliques and a strong level of clique overlap. In CareLab, a larger amount of clique overlap is present, as the results of the hierarchical clustering procedure show, but as only four actors are a member of a clique this only implies a strong core group of four actors, but poor integration on the whole network level. As the Radicalisation network only consists of one clique, obviously no clique overlap occurs in that network. Thus, the Invasive Species case has the strongest integrated network in which all actors actively elaborate upon each other's ideas. Everyone's idea is built upon by at least two other actors given that every actor is a member of a clique.

Actor Importance

Concerning actor importance, we found in the CareLab network that the members of a clique are among the most important actors of the network. Still,

Table 11.2 Network integration 'interaction during meetings' network

	CareLab	Invasive Spcies	Radicalisation
Network density	0.085	0.455	0.156
No. of cliques in network	2	5	1
Average clique size	3	4.6	3
Clique density	0.13	1	0.33
Average clique centrality	0.33	2.56	0.33
Individual clique centralities	0 to 2	1 to 4	0 to 1
Complete integration of cliques at level: (hierarchical clustering of cliques)	2	4	Not applicable, as only one clique is observed in this network
Top important actors (with individual clique centrality)	Local coordinator 1: 93% (clique centrality: 2) Parent 1: 67% (0) Private actor 5: 60% (2) Federal coordinator: 60% (1)	Public actor 2 Flemish: 38% (1) Public actor 2 Walloon: 25% (1) Public actor 1 Brussels: 25% (1)	Federal actor 4 (coordinator) 78% (1) Federal actor 1.67% (0) Federal actor 2.67% (0)

some actors who are regarded as being very important to the process are not included in any clique.

The most important actors in the Invasive Species case are only to some extent a member of a clique. They are only part of one clique, while less important actors are a member of up to four different cliques. We argue that the most important actors are to some extent well integrated in the network, but the network is centralised towards less important actors. In other words, the most important actors do not function as brokers in this network. However, as clique density is high, they are tightly connected to most other actors in the network through clique overlap.

In the successful Radicalisation network, only one clique is observed. The most important actor is present in this clique, however, as no other cliques are present in this network the overall integration of the network is poor. Especially a group of three actors elaborated upon each other's ideas, while only one of these actors was frequently named as very important to the process.

Following these results we see that the Invasive Species case has more or less the same network structure and clique formation in both networks, whereas a clear difference between the networks in the CareLab and Radicalisation cases is observable in terms of clique formation and inclusion of the most important actors. The clique overlap in these latter two cases depends on the type of network, while this is to a lesser extent observable in the Invasive Species case.

DISCUSSION

The goal of this study was to examine how innovative outcomes of collaborative public sector innovation projects can be explained by the network integration of its most important actors. Following from the work on clique overlap and information ties (e.g., Hu et al., 2022; Provan and Sebastian, 1998), collaborative innovation (e.g., Ansell and Torfing, 2014:11) and actor importance (e.g., Cristofoli et al., 2021; Stevens, 2018) we hypothesised that cases with successful innovative outcomes are characterised by a high degree of clique overlap in terms of the most important actors being included in multiple cliques, both during and outside meetings.

We can only confirm our hypothesis for the 'interaction outside meetings' network. Our findings are in line with the argument that clique overlap is related to positive network outcomes. We found that the cases with higher innovative outcomes have a higher integrated network concerning sharing information outside meetings (Cristofoli & Markovic, 2016; Raab et al., 2015). Being a member of multiple cliques in the successful Radicalisation case indicates that the most important actors are in close contact with less important actors and thus act as brokers that connect the different actors with each other. In contract, the few actors in cliques of the less successful CareLab case are among the most important actors, which points to a very important regarded 'in-group' that mainly shares information with each other, while the other less important actors hardly interact with each other. The Invasive Species case shows a high amount of clique density and overlap, however, in this case, the most important actors are less well integrated as they are in fewer cliques.

We expected the same type of integration to be present in the 'interaction during meetings' networks. We expected successful cases to have a higher level of clique density, higher clique centrality of the most important actors, and clique overlap. This is confirmed to some extent as the relatively successful Invasive Species case has a higher level of clique density and more actors are a member of a clique than the CareLab network. However, the successful Radicalisation case follows the pattern of the less successful CareLab case in the 'interaction inside meetings' network: limited cliques, with little involvement of most actors, including the most important ones. We therefore cannot fully confirm our hypothesis for the 'interaction during meetings' network.

CONCLUSION

With this study, we wanted to answer how innovative outcomes of collaborative public sector innovation projects can be explained by the network integration of its most important actors.

Unlike what theory suggests, we have to conclude that successful innovative outcomes are only to some extent explained by the way the most important actors are integrated through clique overlap and depend on the type of network. A well-integrated network (with important actors connected through clique overlap) is not necessarily always a crucial driver for the development of collaborative innovation as the successful Radicalisation case shows poor network integration during meetings. In that sense, the findings refine the argument that information flow is crucial to the development of innovative outcomes (Koliba et al., 2017), as the findings show that a well-connected network is not necessarily always needed. This may indicate that successful cases spend less time on deep discussions and idea generation during broad general meetings and actors in successful cases are more inclined to connect with each other outside meetings to work out details with the relevant actors without the inclusion of non-essential actors. Still, the results also indicate that a certain amount of dispersion of important actors throughout the network seems needed (see also Yi, 2018) as we find that the less successful CareLab case is characterised by an important in-group that does not connect with other actors in the network.

This study builds upon the milestone study concerning clique overlap as the most effective way for information flow in collaborative arrangements (Provan & Sebastian, 1998). Moreover, it was the first to examine clique overlap in combination with the position of the most important actors in the network and to make the distinction between interaction inside and outside meetings. Furthermore, this study is one of the first to examine clique overlap in relation to (digital) public sector innovation. To date, the role of the integration of the most important actors in the networks was only studied to a limited extent, while research suggests that, on the one hand, the network position of certain main actors (such as the coordinators) leads to more effective outcomes (Cristofoli et al., 2021; Raab et al., 2015) and, on the other hand, that actor importance is associated with innovative outcomes (Stevens, 2018). No study had examined the combination of clique overlap as a way of network integration in combination with actor importance. Moreover, usually, no distinction between the complementary networks concerning information sharing inside official meetings and outside official meetings is made.

Our findings suggest that collaborative innovation networks do not necessarily always have to be well integrated through clique overlap both inside and outside official meetings when developing an innovation. The findings are therefore a refinement to the classic studies (e.g., Provan & Sebastian, 1998) that argue that clique overlap is an important driver for positive network outcomes. Clique overlap seems to be of lesser importance for building upon each other's ideas inside meetings and important actors do not necessarily have to engage in interactions in order to achieve successful innovative outcomes.

We do have to acknowledge that the nature of the cases was different. The successful Radicalisation case was working towards a clear end goal, and thus formal meetings were less characterised by idea generation and building upon each other's ideas. Instead, actors interacted with each other outside formal meetings for the arrangement of more practical resources and 'to get things done', which might point to a strong commitment towards innovation and formal meetings were not necessary to let actors interact with each other. In other words, there was no real need to build upon each other's ideas as the end goal and the way to reach it was more or less known already.

In contrast, CareLab was very much in the idea generation phase, so interaction within meetings was necessary to create a process of synergy that enabled the collaborative arrangement to formulate innovative ideas. However, no proper ideas that included all perspectives and that could count on actual support were developed due to poor network integration. Interaction mainly focused on an important 'in-group' and other actors did not interact with each other, thus reducing the process of collective idea generation. For that reason, the phase of the innovation process might explain why poor integration of the network led to a lack of innovative outcomes in CareLab, while this was not an issue in the Radicalisation case. The Invasive Species case shows in that respect a mixture between the two projects. This case was also largely implementation-oriented, however, as some ideas still needed to be decided upon, building upon each other's ideas in formal meetings was still very much necessary. This might be a reason why actors, in this case, are tightly connected both inside and outside meetings, but that lack of active involvement of the most important actors explains why this case does not have the highest innovative outcomes. Practitioners or coordinators of innovation projects should therefore be aware of the phase of the innovation project. They have to determine to what extent integration through clique overlap is needed and when, during or outside meetings, the most important actors should interact more with each other and/or the other actors.

Besides the difference in cases, this study has some other limitations. The data stem from one survey in which respondents were asked to evaluate their own projects. Hence, their opinion on the innovative outcomes might be biased by their experiences in the project. Moreover, it is hard to determine whether importance leads to better integration, or if better integration has led to higher importance. Tsai (2001) and Zhao (2022) argue that central actors are more likely to access useful knowledge from others. Hence, better-integrated actors might be considered more important. In contrast, the resource dependency theory (Hillman et al., 2009) argues that the resources of an actor determine its importance. This study did not examine why certain actors are regarded as being more important, so this is an interesting topic for future research.

Future research should also examine what the ideal network structure is for different types (idea generation-oriented or implementation-oriented) of innovation processes and to what extent our findings are generalisable to other innovation projects. Moreover, the findings indicate that collaborative arrangements aimed at innovation should be aware of the interconnectedness of all actors in the network through clique formation in such a way that important actors are well connected with the others; especially outside official meetings. This study did not examine why actors are more likely to interact with others and thus why certain cliques are formed. Future research should examine what drives interactions in the networks in order to determine how clique formation can be achieved.

NOTES

1. Data were collected in the period March 2017–June 2018.
2. The four items were: (1) No innovative ideas are developed [in this process] … Many innovative ideas are developed [in this project]; (2) The innovative character of [the process] is lower than my initial expectations … The innovative character of the [the process] exceeds my initial expectations; (3) The innovative ideas that are developed in [project name] are not feasible at all … The innovative ideas that are developed in [project name] are very feasible; and (4) The [solutions that have been developed] do not deal with the problems at hand at all … The [solutions that have been developed] truly deal with the problems at hand.
3. Due to a different way of measurement, no network data of non-respondents were available, hence non-respondents in all three cases were excluded from the analysis of the 'elaboration upon other's ideas' network.

REFERENCES

Agger, A., & Lund, D.H. (2017). Collaborative innovation in the public sector – new perspectives on the role of citizens? *Scandinavian Journal of Public Administration, 21*(3), 17–38.

Agger, A., & Sørensen, E. (2018). Managing collaborative innovation in public bureaucracies. *Planning Theory, 17*(1), 53–73.

Agneessens, F., & Wittek, R. (2012). Where do intra-organizational advice relations come from? The role of informal status and social capital in social exchange. *Social Networks, 34*(3), 333–45.

Ansell, C., & Torfing, J. (Eds.) (2014). *Public innovation through collaboration and design*. Routledge.

Ansell, C., Sørensen, E., & Torfing, J. (2021). The COVID-19 pandemic as a game changer for public administration and leadership? The need for robust governance responses to turbulent problems. *Public Management Review*, *23*(7), 949–60.

Bodin, Ö., Sandström, A., & Crona, B. (2017). Collaborative networks for effective ecosystem-based management: A set of working hypotheses. *Policy Studies Journal*, *45*(2), 289–314.

Bommert, B. (2010). Collaborative innovation in the public sector. *International Public Management Review*, *11*(1), 15–33.

Borgatti, S.P., Everett, M.G., & Johnson, J.C. (2013). *Analyzing social networks*. Sage.

Castells, M. (2000). *The rise of the network society: The information age: economy, society and culture* (vol 1, 2nd edn). Blackwell.

Cepiku, D., Giordano, F., Mastrodascio, M., & Wang, W. (2020). What drives network effectiveness? A configurational approach. *Public Management Review*, *23*(10), 1479–503.

Chen, J., Walker, R.M., & Sawhney, M. (2020). Public service innovation: A typology. *Public Management Review*, *22*(11), 1674–95.

Cristofoli, D., & Markovic, J. (2016). How to make public networks really work: A qualitative comparative analysis. *Public Administration*, *94*(1), 89–110.

Cristofoli, D., Trivellato, B., Sancino, A., Maccio, L., & Markovic, J. (2021). Public network leadership and the ties that lead. *Journal of Management and Governance*, *25*(1), 251–74.

Crosby, B.C., 't Hart, P., & Torfing, J. (2017). Public value creation through collaborative innovation. *Public Management Review*, *19*(5), 655–69.

Damanpour, F., Walker, R.M., & Avellaneda, C.N. (2009). Combinative effects of innovation types and organizational performance: A longitudinal study of service organizations. *Journal of Management Studies*, *46*(4), 650–75.

De Vries, H., Bekkers, V., & Tummers, L. (2016). Innovation in the public sector: A systematic review and future research agenda. *Public Administration*, *94*(1), 146–66.

Geuijen, K., Moore, M., Cederquist, A., Ronning, R., & Van Twist, M. (2017). Creating public value in global wicked problems. *Public Management Review*, *19*(5), 621–39.

Gieske, H., George, B., van Meerkerk, I., & van Buuren, A. (2019). Innovating and optimizing in public organizations: Does more become less? *Public Management Review*, *22*(4), 1–23.

Godenhjelm, S., & Johanson, J.E. (2018). The effect of stakeholder inclusion on public sector project innovation. *International Review of Administrative Sciences*, *84*(1), 42–62.

Harary, F., Norman, R.Z., & Cartwright, D. (1965). *Structural models: An introduction to the theory of directed graphs*. Wiley.

Hartley, J., Sørensen, E., & Torfing, J. (2013). Collaborative innovation: A viable alternative to market competition and organizational entrepreneurship. *Public Administration Review*, *73*(6), 821–30.

Hillman, A.J., Withers, M.C., & Collins, B.J. (2009). Resource dependence theory: A review. *Journal of Management*, *35*(6), 1404–27.

Hu, Q., Medina, A., Siciliano, M.D., & Wang, W. (2022). Network structures and network effects across management and policy contexts: A systematic review. *Public Administration*. https://doi.org/10.1111/padm.12835

Jansen, D. (2006). *Einführung in die Netzwerkanalyse: Grundlagen, Methoden, Forschungsbeispiele*. Springer-Verlag.

Kegen, N.V. (2015). Cohesive subgroups in academic networks: Unveiling clique integration of top-level female and male researchers. *Scientometrics, 103*(3), 897–922.
Koebele, E.A. (2019). Policy learning in collaborative environmental governance processes. *Journal of Environmental Policy & Planning, 21*(3), 242–56.
Koliba, C., Wiltshire, S., Scheinert, S., Turner, D., Zia, A., & Campbell, E. (2017). The critical role of information sharing to the value proposition of a food systems network. *Public Management Review, 19*(3), 284–304.
Koppenjan, J.F.M., & Klijn, E.H. (2004). *Managing uncertainties in networks: A network approach to problem solving and decision making*. London: Psychology Press.
Lewis, J.M., Ricard, L.M., & Klijn, E.H. (2018). How innovation drivers, networking and leadership shape public sector innovation capacity. *Revue Internationale des Sciences Administratives, 84*(2), 301–20.
Lopes, A.V., & Farias, J.S. (2020). How can governance support collaborative innovation in the public sector? A systematic review of the literature. *International Review of Administrative Sciences, 88*(1), 114–30.
Lusher, D., Koskinen, J., & Robins, G. (Eds.) (2012). *Exponential random graph models for social networks: Theory, methods, and applications*. Cambridge University Press.
Meijer, A.J. (2014). From hero-innovators to distributed heroism: An in-depth analysis of the role of individuals in public sector innovation. *Public Management Review, 16*(2), 199–216.
Meijer, A., & Bekkers, V. (2015). A metatheory of e-government: Creating some order in a fragmented research field. *Government Information Quarterly, 32*(3), 237–45.
Ngamassi, L., Maitland, C., & Tapia, A.H. (2014). Humanitarian interorganizational information exchange network: How do clique structures impact network effectiveness? *VOLUNTAS: International Journal of Voluntary and Nonprofit Organizations, 25*(6), 1483–508.
Osborne, S.P., & Brown, L. (2011). Innovation, public policy and public services delivery in the UK: The word that would be king? *Public Administration, 89*(4), 1335–50.
Pardo, T.A., & Tayi, G.K. (2007). Interorganizational information integration: A key enabler for digital government. *Government Information Quarterly, 24*(4), 691–715.
Pfeffer, J., & Salancik, G. (1978). *The external control of organizations: A resource dependence perspective*. Harper & Row.
Provan, K.G., & Kenis, P. (2008). Modes of network governance: Structure, management, and effectiveness. *Journal of Public Administration Research and Theory, 18*(2), 229–52.
Provan, K.G., & Milward, H.B. (1995). A preliminary theory of interorganizational network effectiveness: A comparative study of four community mental health systems. *Administrative Science Quarterly, 40*(1), 1–33.
Provan, K.G., & Sebastian, J.G. (1998). Networks within networks: Service link overlap, organizational cliques, and network effectiveness. *Academy of Management Journal, 41*(4), 453–63.
Raab, J., Mannak, R.S., & Cambré, B. (2015). Combining structure, governance, and context: A configurational approach to network effectiveness. *Journal of Public Administration Research and Theory, 25*(2), 479–511.
Saz-Carranza, A., & Ospina, S.M. (2011). The behavioral dimension of governing interorganizational goal-directed networks – managing the unity-diversity tension. *Journal of Public Administration Research and Theory, 21*(2), 327–65.

Scharpf, F.W. (1978). Interorganizational policy studies: Issues, concepts and perspectives. In K.I. Hanf & F.W. Scharpf (Eds.), *Interorganizational policy making* (pp. 345–70). Sage.

Scott, J. (2000). *Social network analysis: A handbook* (2nd edn). Sage.

Seidman, S.B., & Foster, B.L. (1978). A graph-theoretic generalization of the clique concept. *Journal of Mathematical Sociology*, *6*(1), 139–54.

Siddiki, S., Kim, J., & Leach, W.D. (2017). Diversity, trust, and social learning in collaborative governance. *Public Administration Review*, *77*(6), 863–74.

Sørensen, E., & Torfing, J. (2011). Enhancing collaborative innovation in the public sector. *Administration & Society*, *43*(8), 842–68.

Stevens, V. (2018). Managing the ostrich dance: A study on the management of collaborative networks for the promotion of policy innovations. PhD dissertation. University of Antwerp.

Torfing, J. (2019). Collaborative innovation in the public sector: The argument. *Public Management Review*, *21*(1), 1–11.

Tsai, W. (2001). Knowledge transfer in intraorganizational networks: Effects of network position and absorptive capacity on business unit innovation and performance. *Academy of Management Journal*, *44*(5), 996–1004.

Turrini, A., Cristofoli, D., Frosini, F., & Nasi, G. (2010). Networking literature about determinants of network effectiveness. *Public Administration*, *88*(2), 528–50.

Voorberg, W., Bekkers, V., Timeus, K., Tonurist, P., & Tummers, L. (2017). Changing public service delivery: Learning in co-creation. *Policy and Society*, *36*(2), 178–94.

Wegrich, K. (2019). The blind spots of collaborative innovation. *Public Management Review*, *21*(1), 12–20.

Yi, H. (2018). Network structure and governance performance: What makes a difference? *Public Administration Review*, *78*(2), 195–205.

Zhao, J. (2022). Coupling open innovation: Network position, knowledge integration ability, and innovation performance. *Journal of the Knowledge Economy*, *14*(3), 1–21.

PART IV

Conclusions

12. Collaboration for digital transformation: so much more than just technology

Koen Verhoest, Erik Hans Klijn, Lise H. Rykkja and Gerhard Hammerschmid

INTRODUCTION

Responsible governments across the world are constantly searching for new ways to create innovative services in the face of diverse and complex societal challenges and wicked issues. Technological innovations, especially digitalization and the use of different digital tools and components are promising because they can connect and integrate a large variety of services and contexts and improve the accessibility and quality of these services. Collaborating with a variety of stakeholders in this way may result in creative processes and new, improved, and innovative public services as knowledge, resources, and ideas are shared and connected with each other. However, technology is not enough by itself. Notably, governments need to bring actors together and establish know-how on *how to collaborate* to build the necessary digital infrastructure, create innovation, and develop new (digital) services.

The research presented in this book generates evidence from various European countries on the crucial role of collaboration in creating digital transformation and innovative public services. Bringing together government actors and involving external actors, both private sector organizations and users, in the collaboration process is important. When actors with diverse backgrounds collaborate, mutual learning is stimulated, and collective capacity is increased. In this process, new innovative digital solutions that enhance the value of public administrations for citizens, users, and communities can be developed, thereby enabling the public sector to achieve its specific aims.

The digitalization processes that governments currently undertake are transforming the way governments work and interact with citizens and other stakeholders. However, there is also significant heterogeneity in the state of adoption of digital technologies across countries. In many policy sectors, digital transformation through technological innovation can often only be achieved through collaborative partnerships, as no individual government actor alone possesses all the required knowledge, resources, and capabilities to innovate. These partnerships are often public-private collaborations. The public sector or government organizations can direct political and financial incentives, regulatory powers, and public resources towards innovation, while private sector organizations possess specialized knowledge and capabilities to develop them.

Digital transformation is by no means straightforward, rather, it is a complex process of mutual shaping between different actors and technologies, institutions, and economic, political, and socio-cultural factors as described by Miriam Lips in Chapter 2. The political-institutional context within which the transformation happens influences the dynamics and outcomes. Digital transformation implies risks for political decision-makers – risks of failure, excessive cost, and efficiency loss. If successful, it is also linked to considerable efficiency and effectiveness gains. Achievements in digital transformation initiatives in the public sector are not determined by the technology per se, but by deep-seated social, institutional, legal, political, economic, and cultural processes and structures, leading to fragmented and evolutionary outcomes. Digital technology capabilities and socio-cultural, economic, political, institutional, and organizational factors work together.

Collaboration is an essential form of modern governance, especially when addressing the process of government digitalization and reform. It is also a key feature of digital-era governance. Service transformation towards customer-centric ways of public service delivery, for example, often implies the integration of different public services via integrated online portals, the integration of services across different policy domains and government sectors, but also across different government levels. However, collaboration is hard, time-intensive, and requires high levels of administrative, project, and interpersonal skills on the part of leaders. It also requires that the stakeholders are willing and interested in engaging. Without it, collaborations run the risk of being more 'talk' than 'action'.

In this concluding chapter, reflection is given on the chapters and research provided so far. This is done by singling out three main themes: the role of leadership (third section), the conditions for good collaborations (fourth section), and the importance of stakeholder and user involvement (fifth section). Before that, we reflect on the challenges of collaborating for digital transformation (next section). We finish with a section on lessons for practitioners (sixth section) and avenues for further research (seventh section).

CHALLENGES OF COLLABORATION FOR DIGITAL TRANSFORMATION

Undertaking digital projects and transformation initiatives involves substantial challenges and risks. The risk of failure, unclear outcomes, and the significant costs and resources involved are common for such large-scale projects and limit their attractiveness for political leadership. Academic literature points to complexity, risk, and power inequalities as factors that pose challenges to the dynamics and success of collaborative governance, as well as how this is affected by different national politico-administrative contexts (Huxham and Vangen, 2005; Ansell and Gash, 2008; Crosby and Bryson, 2010; Osborne and Brown, 2011; Torfing, 2019). Smartly combining well-designed structures, on the one hand, with matching leadership styles brings the solution to these challenges but is a challenge that we sketch in this section. Strategies for such matching are discussed in the third, fourth, and fifth sections.

Complexity, Risk, and Power Inequalities in Digital Transformation

Complexity is multidimensional and encompasses the specific collaboration process as well as the wider administrative structure, and is frequently differentiated into substantive, strategy, and institutional complexity (Klijn and Koppenjan, 2016). Substantive complexity arises from differences in the perceptions of problems, goals, and envisioned solutions among the various actors involved in a collaboration project. Strategic complexity refers to the varying strategies adopted by different actors within a collaboration to handle tensions and conflicts, whereas institutional complexity reflects the institutional context of formal and informal traditions, policies, laws, and regulations wherein the collaboration is embedded. For digitalization projects issues such as technological complexity or the lack of digital skills and information asymmetries and technological dependencies and legacies add to these complexities (Neumann et al., 2019).

The perception of the risks associated with digital projects (risk of failure, working in unregulated new territories, high financial investments, or unclear accountability arrangements) also substantially affects collaborative dynamics. Risk aversion tends to be higher in public sector organizations, but risk perception is also contingent on the organization's legacy of previous reforms and its capacities to manage risk. Engaging in collaborative projects also might imply the perception of risks such as losing control and autonomy, legitimacy, and resources. Finally, power imbalances within a collaborative relationship can also hamper the collaborative process. Together, these challenges are argued to have an important and oscillating influence on collaborative outcomes.

National Context Matters for Digital Transformation

The relevance of national political-institutional contexts in shaping public management practices and reform trajectories is a central finding of comparative public administration research (Verhoest, 2010; Hammerschmid et al., 2016; Pollitt and Bouckaert, 2017). This can also be observed in the digital transformation of European national governments: The European Commission with its Digital Economy and Society Index[1] and eGovernment Benchmark[2] shows that there are substantial national differences in how digital public services are progressing. For collaborative approaches, the broader national context in which the collaboration takes place similarly plays an essential role in shaping the challenges, dynamics, and solutions that emerge, and the outcome and success of such efforts (Dawes and Pardo, 2002; Emerson, Nabatchi, and Balogh, 2012; Lewis, Ricard, and Klijn, 2018).

National Context Affects How Challenges Are Perceived and Handled

We are able to show how administrative traditions and collaborative history impact the challenges perceived in intergovernmental collaboration in government-wide platforms for digital services. The national institutional environment played an important role in how the relevance of collaboration challenges was perceived and how decisions were made. In the UK stronger power imbalances were found. There were substantial similarities between the German and the Belgium case regarding the complexity challenge, exemplifying a Continental tradition. Another similarity was found between Estonia and Denmark where a combination of higher professionalization, pragmatism, and informality led to a lower perception of complexity. In contrast, the silozation and decentralization present in Germany and Belgium resulted in power games as well as perceptions of strategic complexity and goal conflicts among the stakeholders. The connection between administrative traditions and risks perceived as collaboration challenges were more ambiguous, but also quite similar in the UK, Danish and Estonian case. Interestingly, the impact of administrative traditions on the perception of challenges did not directly lead to the emergence of similar ways to manage and govern the projects. Chapters 5 and 6 show how the project dynamics were contingent on pre-existing structural ties, referred to as the collaborative history that dictated how challenges were established and addressed, making them not only context- but also path-dependent. The structure also reflects system context, as laws, regulations, and previous collaborative management experience often determine joint project design. Thus, project structure and dynamics are highly contextual, making the organizational environment of projects critical when structuring collaboration for digital transformation.

Hence, the relevance of national institutional environments and cultures remains a prominent factor in shaping the way challenges are perceived and handled, and in shaping the dynamics of intergovernmental and public-private collaboration. This also points to the need for further research and research designs that more explicitly test this relationship.

The Challenge of Coordinating Necessary Collaboration by Smartly Combining Design and Leadership

The most common approach to address the challenges that digital platforms face is the establishment of a central coordinating body managing the projects and relationships among the various actors involved. In the smart city collaboration efforts studied in this book, the findings reveal variations in the intensity of challenges and in management interventions over time. The initial formation phase and the early phases of partnerships are characterized by a higher salience of such complexities, interdependencies, and conflicting interests, and require more intense and resource-intensive collaboration. However, they are also superseded by periods of lower intensity. Collaboration efforts can therefore be very resource-intensive and difficult to sustain over a long period.

Structure helps to address these collaboration challenges. Clearly, articulated rules, legitimized communication channels, and established protocols can reduce partners' negative risk perceptions when uncertainty is great and unfavourable strategic turns are likely, especially in opportunistic or conflict-laden settings. Structure can balance power inequalities by assigning clear responsibilities and empowering the weaker side or can mitigate complexity by dividing the project into manageable steps as shown in Chapters 3 and 5. Structure underlying collaboration can be vital to successful digital project execution, but only when implemented by strong leadership. Structures can constrain leadership, but leadership styles can shape structures if they produce undesired outcomes, prove ineffective, or stall project progress (Chapter 3). Chapter 7 notes that flat and informal structures are associated with more collaborative leadership styles, and highly bureaucratic structures require elements of transactional leadership, especially when promoting stability, standardization, and compliance with those structures.

The chapters on intergovernmental collaboration underlined that adaptive and hybrid structures are required to accommodate the unique conditions of intergovernmental collaboration for digital transformation. Chapters 5 and 6 discuss a mix of more traditional, top-down modes and more networked governance approaches, such as in the form of centralized coordination and decentralized management and operations. Given the complexity of the projects, many intergovernmental collaborations maintained a vertical hierarchy. Both in the national platform and smart city cases, the dynamics and challenges were handled primarily through a central coordinating body that held most of the steering power of the projects. They must align with targeted leadership actions and the context in which they are embedded. With these insights, governments can create effective structures for cross-cutting collaboration to deftly ride the wave of public sector digital reform, ultimately fostering synergies and generating public value for governments.

The findings also point to both challenges of public-private collaboration and to its potential for collaborative innovation. The inherent tension between creative ideation and collaborative ability, and inefficiencies and transaction costs surrounding collaboration, is important. Managing the complexities of partnerships implies the investment of substantial time and resources into network activities, resulting in a lowering of performance compared to other organizational arrangements. Successful collaborations require a delicate balance of controlling and managing the collaborative innovation process, while at the same time allowing variance and creativity.

MIXING LEADERSHIP STYLES? LOOKING AT DIGITAL TRANSFORMATION FROM A LEADERSHIP PERSPECTIVE

In the research project TROPICO[3], which was the base for this book, leadership was one of the core perspectives used to look at the collaboration in these digitalization processes. As Lips shows in Chapter 2, research and literature on digital transformation in the public sector, so far, underlines the importance of strong leadership support, both from the political and top civil service levels. Effective leaders of digital government initiatives are described as collaborative or facilitative leaders, bringing people, organizations, and stakeholders together and facilitating collaboration. So, what does research in this book teach us about leadership and how does it relate to the available literature on leadership? This section tries to answer that question by first shortly reviewing the leadership literature and then confronting this with the main findings in the TROPICO research and the findings of this book.

From Top-Down and Transactional Leadership to Facilitation and Collaborative Leadership

The literature on leadership is vast, and the number of leadership styles that have been conceptualized and empirically measured over time is enormous (van Wart, 2012; Zehndorfer, 2014). Over the last decades, the leadership literature has been strongly dominated by the distinction between transformational and transactional leadership (Bass, 1985; Avolio, Waldman, and Yammarino, 1991; van Wart, 2012). In this literature, leadership is not only top-down but also transactional in nature, emphasizing the reward structure in the relationship between the leader and the employee/follower (van Wart, 2012). Leaders motivate people in organizations by payments and other forms of reward systems (Zehndorfer, 2014). Monitoring the performance of employees is therefore important. Parts of the literature argue that transactional leadership is most suited for what they call 'normal performance' (Avolio, Waldman, and Yammarino, 1991; Howell and Avolio, 1993). Here, transactional leadership is something of a 'baseline' to which every other form of leadership is contrasted and compared and a form of leadership that must be complemented by other forms, especially when aimed at innovation (Avolio, Waldman, and Yammarino, 1991; van Wart, 2013). To formulate it more boldly: Transactional leadership is good for taking care of business, but not for achieving innovation. From that understanding, we should see very different styles of leadership in our empirical material than transactional leadership.

The concept of transformational leadership was dominant in the leadership literature from the 1980s (Bass, 1985), but more criticism and alternatives have arrived over the last decade. Transformational leadership emphasizes the charismatic characteristics of leaders. The baseline is that leaders need to change the organization, and the people in it, to achieve necessary (innovative) goals (Bass, 1985; Tichy and Devanna, 1990). The perspective strongly stresses that leaders must recognize the need for change and innovation. For that, they must formulate visions and motivate employees to implement them (Bass, 1985). Like the transactional perspective, the transformational leadership literature puts the leader at the centre of most development and interactions in the organization. But contrary to the transactional perspective there is also a supportive element (Bass, 1985; van Wart, 2012).

The transformational leadership conceptualization has recently been heavily criticized for having a flawed conceptualization and research (Knippenberg and Sitkin, 2013). This links to development over the last two decades, where leadership literature has witnessed two important changes (van Dierendonck, 2010; van Wart, 2012; Ricard et al., 2016). Alternative leadership theories have emerged, and there has been growing attention on interpersonal leadership theories that emphasize how leaders interact with or support employees. Especially theories that emphasize authenticity and stewardship have emerged. The leader is not the centre of the universe anymore, but someone who facilitates employees, builds relationships with them, and empowers them (van Dierendonck, 2010). Contrary to transformational leadership, which points to charisma, this type of leadership is more about building trust. Innovation then comes not so much from the charisma of the leader, but from the quality of the employees and how leaders are able to enhance that. This is also emphasized by leadership perspectives that focus strongly on the ethical character of leadership (see van Wart, 2012).

Another important development is the emergence of theories of collaborative and network leadership (Kickert, Klijn, and Koppenjan, 1997; Agranoff and McGuire, 2001; Ansell and Gash, 2008). There is a broad consensus in the contemporary governance literature that (network) management or collaborative leadership is essential. This type of leadership and/or management is necessary in networks and collaborative settings and is very different from classical images of organizational leadership (Gage and Mandell, 1990; Kickert, Klijn, and Koppenjan, 1997; Agranoff and McGuire, 2001; Huxham and Vangen, 2005). Ansell and Gash (2008) speak of facilitating leadership, where a leader's task is to mediate between actors and empower the collaboration process. Leadership and management strategies are suitable in network and partnership settings and are facilitating, activating actors, activating necessary resources, and enhancing collaboration (Gage and Mandell, 1990; Agranoff and McGuire, 2001; Huxham and Vangen, 2005). A leader in this perspective is a person who carefully examines the network of available actors, connects them to each other, facilitates the exploration of solutions to address problems, and engages the involved actors to deploy the resources needed for implementation (Klijn, Steijn, and Edelenbos, 2010). Consequently, they must build trust and cooperation among actors with different perceptions of the problems in question, different ideas about the most desirable solutions to them, and different interests (McGuire and Agranoff, 2011; Emerson and Nabatchi, 2015; Klijn and Koppenjan, 2016). Innovations, in this view, are achieved by collaborative leaders who connect actors and necessary information and can share their success with others (Torfing, 2019).

Back to Our Empirical Results

The empirical research presented in this book reveals interesting findings that fit well with the recent development of leadership literature discussed above. A main observation in all the chapters in this book is that all the intergovernmental and inter-departmental collaborations showed that effective leadership was key to the progress and success of the digital initiative. But the chapters also have in common that they show that more traditional transactional leadership styles were combined with more collaborative styles of leadership. Hammerschmid, Breaugh, and Racwitz, for instance (Chapter 3), argue that leadership in the context of collaborative digital initiatives is not about an either-or decision of the right leadership style, but instead requires a multifaceted approach that corresponds to changing and sometimes ambiguous project dynamics and needs. In the smart city initiatives, a collaborative leadership style emerged alongside a classical reliance on formal top-down structures and leadership forms, especially where matters were situated in the classical public bureaucracy. A form of layering emerges, where new forms of organization and leadership that rely more on interaction and horizontal relations are added to the classical more formal top-down bureaucratic (and transactional) forms. The same phenomena can be observed when leadership activities are analysed in the five countries we have included. The respondents emphasize contingent leadership, mixing various styles and combining actions that fit in a transactional leadership style with elements that fit in a collaborative style. The mix depends on a variety of project-related factors and differs between countries and contexts. The transactional leadership style is more common in Continental Europe, such as Belgium and Germany, and in the digital platform cases – where central government with a stronger prevalence of hierarchy and siloed structures plays a stronger role. In the smart city cases, collaborative leadership is more prevalent. Another finding relates to a temporal variation of leadership styles especially regarding project phases. In many cases, interpersonal aspects and collaborative leadership in the project initiation and the network-building phase were paramount, whereas in the later phases – which required technical scaling, project implementation, and compliance – transactional leadership styles gained in dominance.

Similar observations about the crucial role of leadership were made when it comes to public-private collaboration. In Chapter 8, Callens et al. point to different clusters of factors that are important for the success of health partnerships with management activities being a main one of them. Leadership/management emerges as a very important condition in the qualitative comparative analysis (QCA) by Callens and Klijn in Chapter 9. Another similar trend is the mixing of styles of leadership. The research presented in the second part of the book clearly emphasizes flexibility and leadership that encourages exploration but also emphasize focus on desired outcomes. In collaborative partnerships, variation (i.e., diversity of perspectives, skills, knowledge, etc.) is an important factor that stimulates learning processes and is crucial for creating innovation. Leadership that encourages the exploration and usage of this diversity, therefore, enhances the innovation process. However, variation also causes complexities as the presence of diverse actors makes the innovation process more difficult. Leadership capable of controlling these complexities and streamlining the innovation process towards desired outcomes is, therefore, essential. Our results show that a proper balance between leadership with an orientation on results and leadership which is flexible and can adapt to the complexity of these collaborative processes is important to create innovative services.

The importance of leadership legitimacy in the empirical chapters is also interesting. This is taken for granted both in the classical transactional, but also transformational leadership literature, where it usually is seen to result from clear authority lines in public bureaucratic organizations. However, the observation of the importance of the legitimacy of leadership in the empirical findings clearly fits more with interpersonal and ethical perspectives on leadership and even more in collaborative forms of leadership that have emerged more recently. New intra-organizational leadership styles, like servant leadership and ethical forms of leadership, also emphasize legitimacy within the organization and of the employees as crucial. Furthermore, network management or collaborative leadership is rooted in the acceptance of the other involved actors in the network of the network manager/collaborative leader (Emerson and Nabatchi, 2015; Klijn and Koppenjan, 2016).

Thus, our empirical material demonstrates both the development towards more interpersonal/ethical styles of leadership and towards more collaborative leadership, as emphasized in the leadership literature. It also illustrates what we can call 'decentralizing leadership'. The leader is no longer the charismatic central person in all innovation processes but more a facilitator, working to enable people and organizations to work together.

Thus, further research should focus less on one type of leadership and should look more into how leadership styles are combined and under what conditions leaders change their styles.

DESIGNING AND STRUCTURING PARTNERSHIPS: THE ART OF COLLABORATION

This book points to several important conditions when looking at the functioning of partnerships that could inform designing and running collaborative partnerships for digital transformation and innovation. Five important conditions can be highlighted:

a. Diversity and size, and how the governance structure should be aligned with these two aspects
b. The interaction patterns of the partnerships
c. Structuring intergovernmental collaboration
d. Combining contractual and network management
e. The role of information and communication technology (ICT) to structure the collaboration

These five conditions seem to be dependent upon the main orientation of the innovation project, being either an open search for an innovative solution which is then implemented or a more focused development and implementation of a pre-chosen solution. We elaborate on each of the five conditions below.

Diversity and Size

The synergy between actors with different knowledge, experiences, perspectives, and resources is considered a crucial advantage of collaborative innovation (Sørensen and Torfing, 2017). There is also a limitation to this: too much diversity can cause fragmentation and tensions. Therefore, actively balancing diversity among the partners to stimulate creative processes, on the one hand, and alignment of perspectives to generate synergy, on the other, is crucial (Nissen, Evald, and Clarke, 2014).

The question of partnership size is very much related to this issue of diversity. The chosen governance structure should be contingent on both size and diversity. In Chapter 8, Callens et al. find that the size and governance structure determine and facilitate the involvement of crucial actors in the collaboration. Governance structures play a crucial role in managing and controlling partnerships, as well as facilitating interaction among stakeholders. These structures define the various responsibilities and roles of the partners and establish practices for decision-making and communication among the different participants. According to the existing literature on network governance modes (Provan and Kenis, 2008), the size of a partnership is typically associated with the type of governance structure employed. Larger partnerships tend to benefit from more formal and centralized governance structures, such as network-administrative organizations or partnerships governed by a lead organization. However, our research on collaborative innovation partnerships reveals a different perspective. In the case of small government-coordinated partnerships with contractual arrangements (e.g., when private partners have been selected through a procurement process), centralized governance by a lead organization proves advantageous. On the other hand, we found that large contractual partnerships for collaborative innovation, which are coordinated by societal actors, benefited from the implementation of distributed governance. This approach emphasizes joint decision-making and equal responsibilities among all partners involved.

The Structure of Interaction Patterns

A second key factor in the effectiveness of partnerships lies in the way interactions between actors are structured. The governance structure plays a critical role in shaping these interactions within the partnership. When communication is lacking between different parts of the governance structure, it can lead to difficulties and even parallel decision-making processes. This lack of communication can also erode confidence in the decisions made within the partnership.

The importance of structured interactions among actors for collaborative innovation dynamics is demonstrated in Chapter 11 by Langbroek and

Verhoest. Previous research has shown that well-connected key actors, such as coordinators, contribute to more effective outcomes. Studies by Raab, Mannak, and Cambré (2015) and Cristofoli et al. (2021) support this notion. Additionally, actor importance has been linked to innovative outcomes, as highlighted by Stevens (2018). In terms of information flow within collaborative arrangements, clique overlap is considered the most effective way. Provan and Sebastian (1998) have observed this pattern.

However, the case studies presented in Chapter 11 challenge the notion that a well-integrated network with important actors connected through clique overlap is always the driving force behind collaborative innovation development. This finding suggests that extensive discussions and idea generation during broad general meetings may not be as crucial for successful cases. Instead, successful actors tend to connect and work out details outside of formal meetings, involving only essential actors. Nonetheless, the results also indicate that a certain level of dispersion of important actors is necessary. In one of the cases, the lack of success can be attributed, at least in part, to an influential in-group that fails to establish connections with other actors in the network. These findings refine the argument that information flow is critical for innovative outcomes, as proposed by Koliba et al. (2017). Therefore, the results in Chapter 11 validate existing theories and empirical findings regarding the importance of information flow, while also adding the insight that clique overlap is not always essential.

Network and Contract Management

The choice of managerial activity holds significant importance, particularly when considering partnerships, collaborative governance, and network governance. Existing literature presents two key ideas regarding this matter. The first idea, rooted in economic-oriented literature, highlights the risk of opportunistic behaviour. It emphasizes the contract as a crucial instrument for managing such projects (Jensen and Meckling, 1976). Accordingly, contract management focuses on monitoring performance based on agreed-upon criteria, adhering to the project timeline, and utilizing budget penalties and sanction mechanisms to ensure collaboration, achieve outcomes, and foster innovation. Consequently, innovation should be explicitly specified in the contract.

In contrast, collaborative-oriented perspectives found in the literature on network and collaborative governance stress that these projects primarily entail collaboration and should be viewed as such. These perspectives underscore the importance of collaborative or network management structures (Agranoff and McGuire, 2001; Huxham and Vangen, 2005; Steijn, Klijn, and Edelenbos, 2011; Emerson and Nabatchi, 2015). They argue that contracts cannot account for all unforeseen events and partner behaviours. In other words, contracts can never be comprehensive. Even extensive information gathering would not suffice to handle unforeseen dynamics and changes over time. Therefore, achieving desired outcomes and innovative results necessitates extensive interaction between partners, along with effective management of these interactions, as emphasized in this book. Various terms are used to describe these management activities (e.g., collaborative governance, network management), but they generally encompass similar strategies.

Consequently, both contract management and network management can coexist within the same partnership. Contract management primarily focuses on the input and output aspects of the collaborative innovation process. For instance, it involves engaging innovation-oriented contractors, incentivizing innovation through contract provisions, and specifying output requirements. A contract-like agreement facilitates the transparent allocation of responsibilities, accountability, resources, and risks. Conversely, network management centres around process-related aspects, such as exploring ideas and perspectives, and facilitating interactions between actors. Recent empirical research supports the combination of different strategies (Warsen, Klijn, and Koppenjan, 2019; Callens, Verhoest, and Boon, 2022).

This trend is also evident in this book, which highlights the significance of contract management for digital transformation projects, particularly due to their complexity, cost, and risk. In the examined collaborative innovation cases, contracts were employed for various purposes, including clarifying interdependencies and partner roles, integrating additional knowledge into the partnership, and preventing conflicts by defining accountability relationships. Contracts also help mitigate the risks associated with participating in a partnership. Our research reveals that collaboration coordinators need to strike a balance between contract rigidity (clear goals and incentives) and contractual flexibility (room for experimentation and adjustments after contract conclusion). Additional process rules, incorporated as part of network management, can aid in achieving this balance.

The research presented in this book emphasizes the crucial role of network management in collaborations for digital transformation, both at governmental levels and between public and private actors. Different strategies of network management, such as connecting, exploring, arranging, and issuing process rules (Klijn, Steijn, and Edelenbos, 2010), are combined in varying ways across different projects. This underscores that effective network management is more of a craft than a set of mere techniques.

The combination of contract management and network management proves beneficial when seeking innovative solutions, including in the context of digital transformation. Specifically, specific combinations of contract management (particularly clear output specifications) and network management (specifically the exploring strategy) yield highly innovative eHealth services.

Our findings reinforce the relevance of integrating relational management strategies with contractual foundations for successful and innovative collaborations (Warsen, Klijn, and Koppenjan, 2019; Callens, Verhoest, and Boon, 2022). Although these findings may relate to the specific context of eHealth innovations, the discovery that a blend of contractual elements and exploring elements generates highly innovative digitalization projects is surprising.

The Use of ICT (Tools) to Structure Collaboration in Partnerships

ICT played a crucial role in the eHealth innovation projects examined in this book. However, its significance extended beyond merely being a technological component. ICT served as a facilitator for effective collaboration among partners in the pursuit of solutions. It played a pivotal role in the collaborative process itself, enabling partners to interact with users through tools like mock-ups and testing platforms. Communication technologies such as Skype, MS Teams, online interaction platforms, and cloud databases like SharePoint and Dropbox were extensively employed to connect diverse partners and coordinate their efforts. Furthermore, certain ICT tools possessed analytical capabilities that supported decision-making. The existing ICT infrastructure also held significant importance.

ICT tools are indispensable for fostering collaboration between partners. They improve interactions and help overcome practical communication barriers. Our research highlighted the vital role of ICT as an enabler of user involvement. Through digital environments created by ICT tools, users can test ideas (e.g., simulations, eHealth tool demonstrations). ICT aids in visualizing and structuring innovative concepts, generating support, and understanding. It facilitates data sharing and enables mutual learning by combining and connecting information and knowledge through databases. These findings suggest that ICT stimulates innovation creation in contexts characterized by high levels of trust between partners. One possible explanation is that ICT reduces the necessity for face-to-face interactions, which are typically required to build trust. Therefore, trust must already be present for ICT to exert positive effects on innovation creation.

The Orientation of the Partnership: Innovative Search or Focused Development?

The optimal design choices for size, diversity, governance structure, and the integration of contract management and network management should be considered as being contingent upon the nature of the innovation project. Specifically, it is important to consider whether the project involves an open search for innovative solutions followed by implementation or a focused development and implementation of a predetermined set of solutions.

Based on our findings, we suggest that in projects characterized by an 'open search' approach, success is more likely to be achieved through large networks comprising diverse actors. These networks should adopt a distributed governance structure with a predominant focus on network management strategies. On the other hand, in projects that involve a narrower range of preselected solutions, it may be more functional to establish smaller collaborations with reduced diversity. These collaborations should be governed in a more centralized manner, employing formal contracts with clear stipulations, incentives, and a combination of network management strategies.

Similarly, in digitalization projects aiming for radical innovations, it is crucial to maintain stability in governance structure, network size, diversity, and management over time, particularly when transitioning from an open search for alternative solutions. Practitioners and project coordinators should be mindful of the project phase in order to determine the extent to which integration through clique overlap is necessary. They should also consider when it is most appropriate for the key actors to interact more intensively with each other and/or with other actors, whether during or outside formal meetings. Exploring these temporal dynamics requires further research.

POSSIBILITIES AND PITFALLS OF INVOLVING STAKEHOLDERS AND USERS

Expertise, Position, and Drivers of Private Partners and Stakeholders Matter

As shown in Chapter 9, the expertise, positions, and drivers of the individuals and organizations shape the partnerships and their outcomes. The individuals and organizations from the private and non-profit sectors involved in the collaborative innovation partnerships bring specific expertise and skills. These include ICT expertise, legal expertise, sector-specific expertise, and technical expertise. ICT knowledge is particularly important, while legal expertise is necessary for contract drafting and dealing with data protection. Technical expertise is required for addressing specific issues related to the innovation process. Also, the positions of the actors in the policy sector influence the partnerships. Some actors have access to resources through their positions, such as relevant service users or political support. Actors responsible for ICT infrastructure are involved to ensure interoperability. Their influence is significant, as they control essential data exchange infrastructure. Moreover, various drivers motivate the actors to participate. Some seek innovation to reduce costs or improve services. Others aim to solve urgent problems affecting them directly or indirectly. Opportunities to develop existing services or access economic benefits also drive participation from private actors. Private actors see the projects as opportunities to expand their market shares and gain new knowledge and user experiences. Our research shows it is crucial to select partners purposefully on their expertise, sector position, and their drivers to foster collaboration for digital transformation.

The Extent, Timing, and Perspectives of Involving Users Matter

The research presented in this book highlights also the importance of engaging end users in digitalization projects. Service users play a crucial role as key stakeholders in collaborative innovation processes for digital transformation. They possess valuable insights into the effectiveness and relevance of digital tools and services. Involving them is widely recognized in the literature as a major catalyst for successful innovations (Cinar, Trott, and Simms, 2019).

In Chapter 4, Callens and Verhoest identify several reasons for involving users. Firstly, users can provide legitimacy to the innovation process and its outcomes. They bring valuable information about their expectations and demands, which shapes the development of innovations (Sørensen and Torfing, 2017). Secondly, involving users allows partnerships to access 'sticky information' related to service experiences and local implementation contacts (von Hippel, 1994; Simmons and Brennan, 2017). This information can be used to customize the service to meet the diverse needs of users. Importantly, users do not have strategic motives to keep innovations hidden from competitors, making them more open to participation (Roszkowska-Menkes, 2017).

While existing literature has extensively explored user participation in policy and service creation (e.g., Pestoff, 2014; Brandsen and Honingh, 2016; Nabatchi, Sancino, and Sicilia, 2017), and scholars suggest that user involvement in collaborative partnerships stimulates innovation processes (e.g., 'coproduction for innovation' by Nesti, 2018; 'quadruple helix' innovation by Carayannis and Campbell, 2009), there is still much to learn about the specific conditions under which user involvement leads to collaborative service innovation (though, see Callens, 2022). Some chapters in this book provide more insights into some of these conditions.

To maximize the potential of user involvement in digital innovation projects, three dimensions of user involvement are crucial: timing, intensity, and alignment among partners on how users should be involved and supported by the partnerships.

In the case studies discussed, the timing of user involvement played a crucial role (Alam, 2002). Users can contribute valuable insights throughout various stages of the innovation process, including problem definition, idea generation, testing, implementation, and evaluation. Their involvement can occur in one or more of these stages. When users are engaged early in the innovation process, there are greater opportunities to integrate their input into the final innovation, as observed in partnerships that resulted in highly innovative services. Additionally, incorporating users' perspectives during the ideation phase ensures that ideas directly address their concerns.

Another key aspect of successful user engagement relates to the intensity of their involvement, which refers to the extent to which their input is considered in decision-making processes (Alam, 2002). Users can contribute in different ways, such as providing advice, co-producing with the partnership, or even leading certain aspects of the innovation process (Arnkil et al., 2010; Holgersson and Karlsson, 2014). Empowering users by involving them intensively has been associated with improved service quality, while the absence of empowered users is seen as a significant barrier to public service innovation (Voorberg, Bekkers, and Tummers, 2015; Cinar, Trott, and Simms, 2019). In partnerships where users play a crucial role in decision-making, their intensive involvement becomes particularly important during the conceptual, testing, and implementation phases of the innovation process. The cases examined in this book highlight the need to go beyond mere user information provision and instead include user input in decision-making or grant users decision-making authority. Establishing a governance structure that assigns users a specific role can facilitate extensive user involvement. Moreover, collaboration partners should create a learning environment that enables the open sharing of feedback and experiences with users.

Our research indicated that collaboration partners may have different viewpoints on how to involve users in the innovation process in terms of their motivations, timing, intensity, and the way the partnership can support their involvement. In Chapter 11, we examine the partners' viewpoints on user involvement in collaborative innovation projects. Specifically, we explore four roles of user involvement: users as legitimators, customers, partners, and self-organizers. To gather these perspectives, we employ Q-methodology, surveying 50 partners engaged in eHealth collaborations for service innovation.

The findings reveal that the partners' views on user involvement lie somewhere between perceiving users as 'customers' within a New Public Management paradigm and as partners within a network collaboration and co-production perspective associated with New Public Governance. Two distinct empirical profiles of user involvement emerge from the analysis. The first profile is output-oriented, where user involvement is seen as a means to achieve a satisfactory end product. The second profile is process-oriented, considering users as active participants in the collaborative dynamics of the innovation process. However, our study shows also that partners in collaborative innovation partnerships generally do not view users as initiators or leaders of the innovation process. This limitation suggests a reluctance to incorporate self-governance structures in user involvement. This hesitance may stem from the advanced knowledge requirements in ICT and health necessary for the development of eHealth innovations, as well as the complexity of the healthcare sector. Additionally, it is important to acknowledge that users themselves may hold different expectations and visions regarding their involvement and the partnership's role, which can differ from the partners' expectations (see Callens et al., 2023).

Managing these differences in expectations is crucial to effectively facilitate user involvement. Variances in partners' viewpoints on user involvement can lead to role-related conflicts, both among partners and between partners and users. Consensus-building becomes more challenging in such cases and may require strong project leaders. Therefore, when assembling the partnership, the project coordinator should consider the diverse users needed based on the partners' vision to ensure alignment and avoid any miscommunication regarding user involvement.

LESSONS FOR POLICY AND PRACTICE

There are numerous recommendations to be drawn from the insights in this book, both for policymakers seeking to enhance policies for digital transformation, and for those directly involved in public-private collaborations engaging in the digital transformation of government and society. Overall, public sector actors that are ready to embrace collaboration and ensure that both rhetoric and practice go hand in hand can make a considerable contribution towards the digital transformation of governments. Technology is often seen as an innovation goal per se, and in the context of digitalization, governments often focus on mobilizing IT/digital skills and expertise. However, as highlighted in this book, one should also pay close attention to the development of collaboration, and related skills and competencies, and one should keep in mind that government officials are often motivated less by the innovative technology itself than by the associated public value.

Collaboration means sharing and connecting knowledge, resources, and ideas. The public sector's collaboration with other stakeholders, that is, private sector organizations and users, can, as our research has shown, result in mutual learning, creative processes, and new, improved, and innovative public services. Selecting actors with the needed knowledge, skills, resources, drivers, and incentives is essential. Identifying the appropriate stakeholders and including the right mix of actors can, furthermore, prevent distrust and conflict, enhance creative ideation processes, and secure support for newly created services. A diversity of perspectives, skills, and knowledge tends to stimulate learning processes, and different types of expertise on ICT, technical issues, and legal knowledge can drive the innovation process forward. Actor diversity allows for diversification, synergies, and creativity in the innovation process. New associations between distinct ideas and perspectives are formed when individuals recombine ideas and build on each other's knowledge.

Processes that stimulate the exchange of ideas and knowledge facilitate mutual learning and can increase the collective capacity to develop new solutions. Conscious consensus-building between the collaborating partners helps to select desirable ideas and stimulates convergence towards a shared solution. Furthermore, actively building commitment towards implementing the solution ensures that the partners mobilize resources towards this end. Providing sufficient incentives, visualizing past achievements and opportunities, and continuously communicating the benefits that collaboration brings in the context of government digitalization will further cultivate a motivation to work together. At the same time, collaborating in the digital era means working in a constantly evolving context, shaped by new technologies, complex projects, and high risks regarding budget and project outcomes.

Designing an appropriate governance structure, fitting both the size and type of collaboration, and encouraging creative processes as well as goal alignment, is crucial. Collaboration often needs a fundamental agreement between the involved partners and is more likely when a shared vision and an agreement on central goals have been reached. At least some degree of formal rules on how to act in the collaborative arrangement is helpful here. Formalization may enhance the predictability of collaboration and increase the actors' willingness to engage in the process beyond the initial stages. Specifying the demands, mutual expectations, costs, and risks related to the collaboration is likely to enhance performance. Identifying clearly the desired outcome helps to align potentially diverse goals and objectives. Paying attention to the development of collaboration and networking skills within the collaboration can strengthen the capacity to communicate, create shared meaning, resolve conflicts, and overcome resistance to change.

Contract management can help to connect partners and tighten engagement and commitment to the project. However, formal agreements should also aim to avoid stifling creativity and flexibility. They can reduce risks and uncertainties and increase accountability between partners, but sufficient design freedom should also be preserved by limiting restrictions on creative solutions. Size is an important factor in this context. In small government-coordinated and contractual partnerships, governance by a lead organization is advantageous. Large contractual partnerships coordinated by societal actors, however, will benefit more from governance arrangements with a focus on joint decision-making and equal responsibilities among the partners.

Thus, proper management ensures that the collaboration benefits from the full added value of each individual partner. Managing the interactions between the partners ensures goal alignment, trust, mutual learning, and transparent communication. As seen in many of our case studies, combining different management strategies and leadership styles can be beneficial. Hence, the coordinator or leader should make sure to explore the ideas and perspectives of all partners, connect their resources and interests, introduce governance structures that stimulate interactions, and apply process rules that clarify how the collaboration will operate. Leadership that encourages the exploration and usage of diversity may enhance the innovation process. However, diversity also causes complexities. Collaboration between actors from diverse backgrounds generates innovative ideas but might also cause tensions. Effective network and conflict management are therefore important. Incentives and drivers related to the content of the project, as well as economic interests, are important motivators. Leadership capable of controlling these complexities and streamlining the innovation process towards the desired outcomes is essential.

Encouraging the collaborating actors to be open to the ideas of others and building a learning environment where new knowledge and ideas can emerge facilitates the innovation process. This is stimulated by feedback processes concerning ideas and perspectives, experimentation, trial-and-error behaviour, and by introducing new (external) knowledge in the partnership. Partners' capacity to connect with others, connect ideas, learn, think creatively, and be consensus-oriented is valuable. Strong organizational support (e.g., a clear mandate, dedicated time, and specialized training) will furthermore increase commitment to the project but should also provide freedom and autonomy to encourage interactions and out-of-the-box thinking.

Trust is a key condition for successful collaboration. It emerges from interpersonal connections and repeated interactions over time and can be built via formal meetings but also through informal interactions. Encouraging open and transparent two-way communication between the partners and ensuring feedback about the partners' ideas and perspectives stimulates such trust. Together with conscious consensus-building, it helps manage conflicts, increases goal alignment, and in the end, ensures that the actors are willing to spend resources to adopt the innovation.

Actively seeking and securing both internal and external support is important. The collaborating partners need a clear mandate as organizational representatives in the partnership from the higher management. Many of the cases we investigated show that projects that successfully created highly innovative services had secured external support from collaborating organizations, such as elected politicians, the media, and the broader policy sector, before or during the project.

The users of innovative tools and services are crucial assets. Their involvement is most successful when their viewpoints are aligned with the viewpoints of the collaboration partners. Users should therefore be consciously and actively engaged in the innovation process and should be given the possibility to influence both processes and outcomes, for example, through co-creation or co-leadership. Adopting a governance structure in which the users have a specific role is especially helpful. Ideally, they should be involved in all phases of the innovation process – problem definition, idea generation, testing, implementation, and evaluation. Eliminating barriers to their active involvement, for example, removing rules and procedures that hinder collaboration, ensuring reliable information, and avoiding unbalanced representation will help towards this end. Digital tools can also enhance interactions and help overcome communication barriers. Creating digital environments where ideas can be tested (e.g., through simulations or demonstrations) can enable more user involvement. Generating such user feedback stimulates interactions between partners and can be used to improve usability. Digital tools can also help visualize and structure new ideas, thereby generating mutual learning, support, and commitment to new solutions.

Therefore, leadership and process-related conditions are essential for overcoming barriers to interaction and innovation, the empirical results in this book show. The successful leader sets the ground rules, builds trust by upholding these rules, facilitates dialogue, and helps to identify areas of mutual gains. Leaders also play an important role in handling project complexities, navigating power imbalances, and tracking processes and performance. Collaborative leadership focuses on the ability of leaders to engage and collaborate with a heterogeneous team of actors with a diversity of opinions and ideas, building strategies towards common goals. The protagonists of collaboration for digital transformation should make sure to build a shared vision and manage the relational capital of the collaboration, encourage open and transparent communication between the partners, create communication and networking opportunities, stay connected to all relevant project stakeholders in the governance structure, and take the time to build relationships and trust. This type of leadership is important in many collaborations.

However, the changing nature and dynamics of collaboration projects and the multifaceted nature of digitalization projects also imply that we cannot assume this is always the most effective leadership style. Our research also shows that traditional hierarchical, or transactional, leadership remains highly relevant in many cases. This is especially true within the context of large, resource-intensive, and complex digitalization projects. Especially leaders within large, hierarchically organized government organizations need to find the right balance between a transactional leadership style focusing on goals, monitoring, and incentive structures, and a collaborative style aiming at bringing new and different actors together, motivating, developing trust, and facilitating communication. Therefore, the leader of a collaboration should adopt a leadership style that works for the context and stage of the collaboration. More specifically, when aiming at building trust and legitimacy, and developing a shared understanding, a collaborative style focusing on interpersonal relations, facilitating dialogue, bringing actors together, and supporting learning and out-of-the-box thinking should be adopted. However, when aiming at achieving results, gaining stability, and compliance, and in the phase of innovation exploitation and scaling, a transactional leadership style, emphasizing objectives, accountability, monitoring success, and pushing for more time and energy, will be useful.

AVENUES FOR FUTURE RESEARCH

This book has given considerable new insights on many topics related to digital transformation and collaboration. Nevertheless, there is no doubt that there are still many questions that deserve more research attention.

Our research has been more focused on the presence of, and interplay between, conditions, structures, leadership, management and innovation, and digital transformation itself, and less on the mechanisms that resulted in these observations. Future research should therefore examine more – through qualitative research such as process tracing – the actual mechanisms involved in these processes.

It is clear from the findings that different factors, such as complexity, risk, and power imbalance interact with one another and may serve to compensate for or intensify collaborative problems. This also has an impact on the potential for and effects of digital transformation. In Chapters 3 and 7, Hammerschmid, Breaugh, and Rackwitz point out that future research should dig deeper into the dynamics of these challenges in the context of intergovernmental collaboration. Apart from Klijn and Koppenjan's (2016) work on understanding collaborative complexities, few researchers have dug deeper into the different types of risk and power that emerge in collaboration projects. These can, for example, be related to reputation, financial, and/or political issues. If and how they impact collaboration behaviours differently is one area where more knowledge is needed and would provide a more nuanced understanding of collaboration conditions.

Future research should therefore further investigate how and to what degree interactions take place to address these challenges. In doing so, it may be possible to use certain structural arrangements to combat a variety of different challenges and better understand how each unique dynamic (e.g., power and trust) may influence each other. Additional research could also study the impacts that individual system challenges can have on one another, on top of their anticipated impact on the overall outcome.

The chapters for the most part have a within-country or within-case outlook. An examination of how the forces outside a single country impact both the collaborative and digital approaches undertaken in other countries should be conducted. A future research avenue could thus examine the extent to which policy diffusion and policy transfer are occurring across different countries. There are examples of radical openness, for example, through the sharing of open-source digital tools across countries. Whether or not this leads to cross-country adoption of specific systems, or if it is better to develop more grassroot, bespoke solutions would be interesting. Such research can also be tied to the work and influence of international organizations and supranational governing bodies, such as the EU, in encouraging (or mandating) digital transformation.

Callens and Klijn (Chapter 9) emphasize that future research might look beyond the assumptions of economic and governance theory and explore other conditions than contract characteristics and management. The characteristics of the partners, the different phases of the innovation process, or the characteristics connected to the nature of the innovation (e.g., technological sophistication of ICT-enabled service innovations) are likely important. More research into these factors might further unravel the core dynamics of cross-sectoral collaboration for innovation.

The conceptualization of innovation that our approach builds on is more context-dependent and based on the perceptions of the involved actors. However, innovation is also generated through the development of more advanced technologies. Future research could therefore consider extending the concept of innovation to a more context-independent understanding and include elements of technological sophistication. One could analyse the extent of technological sophistication in terms of certain advanced technological components being present/combined or not in collaboration. Our preliminary analyses indicate that conditions that stimulate perceived innovativeness might also affect technological sophistication. An in-depth analysis of which conditions can lead to technologically sophisticated innovations would be another avenue for further research.

Our research also highlights the essential role that politics and political actors play in the implementation of the projects. For many, a lack of political will became a hindrance, even if the goals were admirable. In other cases, too much political backing disrupted the ability for genuine collaborative arrangements to emerge due to power imbalances. In this regard, future research could examine the role and importance of political leadership at different levels of government and how they may become a help or hindrance to the collaborative process.

Another area of research could be to examine the notion of collaborative resistance, and why, when given the choice, some actors or organizations choose not to participate in collaboration projects. It could be related to collaboration dynamics such as risk, power, and complexity, but also to a lack of skills or understanding of digital processes, a lack of motivation due to fear of change, or a lack of people able to engage with the process.

Long-term project maintenance is also an area where we lack knowledge. Potential research questions include aspects such as the dependency of the project managers on the digitalization project, if and how a handover may occur, if and how collaboration changes in the maintenance rather than the set-up phase of a project, the role of collaboration inertia in the long-term sustainability of collaboration networks, and how the challenges and dynamics may change as projects become more mature.

Future studies should also examine the evolutionary aspects of project implementation. Studying projects in their more mature stages would give deeper insight into what skill sets are needed after the initial necessary network structures have been established. More transactional styles of interaction, technical scaling, and compliance might be more important in later stages. This kind of research could contribute to a better understanding of how leadership can help digital transformation succeed in the long term, hopefully yielding improved service provision and public value increase.

NOTES

1. https://digital-strategy.ec.europa.eu/en/policies/desi-digital-public-services
2. https://digital-strategy.ec.europa.eu/en/library/egovernment-benchmark-2022
3. The TROPICO project received funding from the European Union's Horizon 2020 research and innovation programme under grant agreement No. 726840. For more information: https://cordis.europa.eu/project/id/726840

REFERENCES

Agranoff, R., and McGuire, M. (2001). Big questions in public network management research. *Journal of Public Administration Research and Theory*, *11*(3), 295–326.

Alam, I. (2002). An exploratory investigation of user involvement in new service development. *Journal of the Academy of Marketing Science*, *30*(3), 250–61.

Ansell, C., and Gash, A. (2008). Collaborative governance in theory and practice. *Journal of Public Administration Research and Theory*, *18*(4), 543–71.

Arnkil, R., Järvensivu, A., Koski, P., and Piirainen T. (2010). *Exploring quadruple helix outlining user-oriented innovation models*. Final Report on Quadruple Helix Research for the CLIQ project.

Avolio, B.J., Waldman, D.A., and Yammarino, F.J. (1991). Leading into the 1990s: The four I's of transformational leadership. *Journal of European Industrial Training*, *15*(4), 9–16.

Bass, B.M. (1985). *Leadership and performance beyond expectations*. Free Press

Brandsen, T., and Honingh, M. (2016). Distinguishing different types of coproduction: A conceptual analysis based on the classical definitions. *Public Administration Review*, *76*(3), 427–35.

Callens, C. (2022). User involvement as a catalyst for collaborative public service innovation. *Journal of Public Administration Research and Theory*, *33*(2), 329–41.

Callens, C., Verhoest, K., and Boon, J. (2022). Combined effects of procurement and collaboration on innovation in public-private-partnerships: A qualitative comparative analysis of 24 infrastructure projects. *Public Management Review*, *24*(6), 860–81.

Callens, C., Verhoest, K., Klijn, E.H., Nõmmik, S., and Pina, V. (2023). How service users envision their engagement in processes of collaborative innovation: A Q-methodological study on user involvement in eHealth collaborations. Public Policy and Administration, https://doi.org/10.1177/0952076723117

Carayannis, E.G., and Campbell, F.J. (2009). 'Mode 3' and 'Quadruple Helix': Toward a 21st century fractal innovation ecosystem. *Technology Management*, *46*(3–4), 201–34.

Cinar, E., Trott. P., and Simms. C. (2019). A systematic review of barriers to public sector innovation process. *Public Management Review*, *21*(2), 264–90.

Cristofoli, D., Trivellato, B., Sancino, A., Maccio, L., and Markovic, J. (2021). Public network leadership and the ties that lead. *Journal of Management and Governance*, *25*(1), 251–74.

Crosby, B., and Bryson, J. (2010). Integrative leadership and the creation and maintenance of cross-sector collaborations. *Leadership Quarterly*, *21*(2), 211–30.

Dawes, S., and Pardo, T. (2002). Building collaborative digital government systems. In W. McIver and A. Elmagarmid (Eds.), *Advances in digital government* (Vol. 26, pp. 259–73). Springer.

Emerson, K., and Nabatchi, T. (2015). *Collaborative governance regimes*. Georgetown University Press.

Emerson, K., Nabatchi, T., and Balogh, S. (2012). An integrative framework for collaborative governance. *Journal of Public Administration Research and Theory*, *22*(1), 1–29.

Gage, R.W., and Mandell, M.P. (Eds.) (1990). *Strategies for managing intergovernmental policies and networks*. Praeger.

Hammerschmid, G., Van de Walle, S., Andrews, R,. and Bezes, P. (Eds.) (2016). *Public administration reforms in Europe: The view from the top*. Edward Elgar.

Holgersson, J., and Karlsson, F. (2014). Public e-service development: Understanding citizens' conditions for participation. *Government Information Quarterly*, *31*(3), 396–410.

Howell, J.M., and Avolio, B.J. (1993). Transformational leadership, transactional leadership, locus of control, and support for innovation: Key predictors of consolidate d-business-unit-performance. *Journal of Applied Psychology*, *78*(6), 891–902.

Huxham, C., and Vangen, S. (2005). *Managing to collaborat: The theory and practice of collaborative advantage*. Routledge.

Jensen, M.C., and Meckling, W.H. (1976). Theory of the firm: Managerial behavior, agency costs and ownership structure. *Journal of Financial Economics*, *3*(4), 305–60.

Kickert, W.J.M., Klijn, E.H., and Koppenjan, J.F.M. (Eds.) (1997). *Managing complex networks: Strategies for the public sector*. Sage.

Klijn, E.H., and Koppenjan, J.F.M. (2016). *Governance networks in the public sector*. Routledge.

Klijn, E.H., Steijn, B., and Edelenbos, J. (2010). The impact of network management on outcomes in governance networks. *Public Administration*, *88*(4), 1063–82.

Knippenberg, D., and Sitkin, S.B. (2013). A critical assessment of charismatic transformational leadership research: Back to the drawing board? *The Academy of Management Annals*, *7*(1), 1–60.
Koliba, C., Wiltshire, S., Scheinert, S., Turner, D., Zia, A., and Campbell, E. (2017). The critical role of information sharing to the value proposition of a food systems network. *Public Management Review*, *19*(3), 284–304.
Lewis, J., Ricard, L., and Klijn, E. (2018). How innovation drivers, networking and leadership shape public sector innovation capacity. *International Review of Administrative Sciences*, *84*(2), 288–307.
McGuire, M., and Agranoff, R. (2011). The limitations of public management networks. *Public Administration*, *89*(2), 265–84.
Nabatchi, T., Sancino, A., and Sicilia, M. (2017). Varieties of participation in public services: The who, when, and what of coproduction. *Public Administration Review*, *77*(5), 776–76.
Nesti, G. (2018). Co-production for innovation: The urban living lab experience. *Policy and Society*, *38*(3), 310–25.
Neumann, O., Matt, C., Hitz-Gamper, B., Schmidthuber, L., and Stürmer, M. (2019). Joining forces for public value creation? Exploring collaborative innovation in smart city initiatives. *Government Information Quarterly*, *36*(4), 101411.
Nissen, H.A., Evald, M.R., and Clarke, A.H. (2014). Knowledge sharing in heterogeneous teams through collaboration and cooperation: Exemplified through Public–Private-Innovation partnerships. *Industrial Marketing Management*, *43*(3), 473–82.
Osborne, S.P., and Brown, L. (2011). Innovation, public policy and public services delivery in the UK: The word that would be king? *Public Administration*, *89*(4), 1335–50.
Pestoff, V. (2014). Collective action and the sustainability of co-production. *Public Management Review*, *16*(3), 383–401.
Pollitt, C., and Bouckaert, G. (2017). *Public management reform: A comparative analysis – into the age of austerity*. Oxford University Press.
Provan, K., and Kenis, P. (2008). Modes of network governance: Structure, management, and effectiveness. *Journal of Public Administration Research and Theory*, *18*(2), 229–52.
Provan, K.G., and Sebastian, J.G. (1998). Networks within networks: Service link overlap, organizational cliques, and network effectiveness. *Academy of Management Journal*, *41*(4), 453–63.
Raab, J., Mannak, R.S., and Cambré, B. (2015). Combining structure, governance, and context: A configurational approach to network effectiveness. *Journal of Public Administration Research and Theory*, *25*(2), 479–511
Ricard, L.M., Klijn, E.H., Lewis, J.M., and Ysa, T. (2016). Assessing public leadership styles for innovation: A comparison of Copenhagen, Rotterdam and Barcelona. *Public Management Review*, *19*(2), 134–56.
Roszkowska-Menkes, M. (2017). User innovation: State of the art and perspectives for future research. *Journal of Entrepreneurship, Management and Innovation*, *13*(2), 127–54.
Simmons, R., and Brennan, C. (2017). User voice and complaints as drivers of innovation in public services. *Public Management Review*, *19*(8), 1085–104.
Sørensen, E., and Torfing, J. (2017). Metagoverning collaborative innovation in governance networks. *American Review of Public Administration*, *47*(7) 826 –39.

Steijn, A.J., Klijn, E.H., and Edelenbos, J. (2011). Public private partnerships: Added value by organizational form or management? *Public Administration, 89*(4), 1235–52.

Stevens, V. (2018). *Managing the ostrich dance: A study on the management of collaborative networks for the promotion of policy innovations*. PhD dissertation. University of Antwerp. https://repository.uantwerpen.be/docman/irua/dc1bbc/150403.pdf

Tichy, N.M., and Devanna, M.A. (1990). *The transformational leader.* Wiley.

Torfing, J. (2019). Collaborative innovation in the public sector: The argument. *Public Management Review, 21*(1), 1–11.

Voorberg, W.H., Bekkers, V.J.J.M., and Tummers, L.G. (2015). A systematic review of co-creation and co-production: Embarking on the social innovation journey. *Public Management Review, 17*(9), 1333–57.

van Dierendonck, D. (2010). Servant leadership: A review and synthesis. *Journal of Management, 37*(4), 1228–61.

von Hippel, E. (1994). 'Sticky information' and the locus of problem solving: Implications for innovation. *Management Science, 40,* 429–39.

van Wart, M. (2012). *Leadership in public organizations: An introduction.* M.E. Sharpe.

van Wart, M. (2013). Lessons from leadership: The Theory and the contemporary challenges of leaders. *Public Administration Review, 73*, 553–65.

Verhoest, K. (2010). The relevance of culture for NPM. In P. Lægreid and T. Christensen (Eds.), *Ashgate research companion to New Public Management* (pp. 47–64). Ashgate.

Warsen, R., Klijn, E.H., and Koppenjan, J. (2019). Mix and match: How contractual and relational conditions are combined in successful public–private partnerships. *Journal of Public Administration Research and Theory, 29*(3), 375–93.

Zehndorfer, E. (2014). *Leadership: A critical introduction.* Routledge.

Index